Thinking Like a Physicist

Thinking Like a Physicist

Physics Problems for Undergraduates

A collection of problems and solutions, written by the staff of the Physics Department of the University of Bristol, and edited by

N Thompson, PhD, FInstP
Emeritus Professor of Physics
University of Bristol

Adam Hilger, Bristol, Philadelphia and New York

British Library Cataloguing in Publication Data

Thinking like a physicist: a collection of
problems and solutions.
1. Physics — Problems, exercises, etc.
I. Thompson, N. II. University of Bristol.
Department of Physics
530′.076 QC32

ISBN 0-85274-513-3

First printed 1987
Reprinted 1990

Published under the Adam Hilger imprint by IOP Publishing Ltd
Techno House, Redcliffe Way, Bristol BS1 6NX, England
335 East 45th Street, New York, NY 10017-3483 USA
US Editorial Office: 1411 Walnut Street, Philadelphia, PA 19102, US

Typeset by Mathematical Composition Setters Ltd, Salisbury

Introduction

The Objective

For the past quarter of a century it has been the practice in the physics department of the University of Bristol to include a 'General Paper' in the final examination for the honours degree. Its essential characteristics are that it does not refer directly to the subject matter of any particular course of lectures; that, for the most part, the questions do not refer to the more advanced parts of the three-year curriculum—'difficult questions on elementary physics' was the original prescription; and that the questions are all of the problem type, calling for the application of principles which, it is hoped, the students have mastered. The reason for the introduction of the general paper was a wish to escape the criticism—not always unjustified—that success in examinations relied heavily on the reproduction of material that the student had learned. We wished to see whether he or she could also *use* this material. By making the paper a significant part of the final examination we hoped to encourage the cultivation of a group of skills which we regarded as an important constituent of the expertise of the professional physicist. Briefly, these include the ability to convert a 'real' problem into a 'model' that is susceptible of quantitative analysis, by extracting the essential elements; to analyse the behaviour of the model, making whatever approximations are necessary, and to be aware of the consequences of these approximations; to dredge up from his or her store of information items that are relevant to a question presented out of context; to do rough order-of-magnitude calculations and to realise when they could be useful.

Although the character of the paper has undergone changes in detail over the years, the main objectives have remained unaltered. The questions are sometimes classified under two broad headings—the 'well defined' and the 'open-ended'—and each year those who set the paper try to maintain a balance between the two. Although this dichotomy is a convenient concept, there is, in fact, a continuous gradation between the two extremes. A really well defined question has one answer which is right. It will usually call for the development of a mathematical argument not previously encountered by the student in quite the same form, and may combine concepts from different branches of physics, not usually associated. A related—and very important—type of question involves the making of some approximations, either physical or mathematical, in order to obtain an answer which, if it cannot be described as 'right', is at least reasonable. Yet another type will rely centrally on the making of order-of-magnitude estimates, based on experience or memory of related problems; these are the 'back-of-an-envelope' calculations which play such an important role in discussions amongst physicists. Finally, there is the type of question in which the emphasis is on ideas, physical intuition or creative thinking—in the fashionable jargon. There is no 'right' answer; indeed, there may be no answer at all. It was of this type that one of our colleagues once said 'I don't know the answer—but I am sure that I can distinguish a good attempt from a poor one'. Questions beginning 'Write an essay on ...', while not forbidden, are not encouraged. It will be seen that all the types described above, and others in similar vein, call for the deployment of intellectual skills that play only a small part in traditional written degree examinations, yet which are of prime importance in terms of the ability of a physicist to 'do' physics in the real world.

The examination paper that includes these questions usually lasts for three hours. A wide choice of questions is always given, but the rubric has varied considerably over the years. 'Good marks can be obtained for answers to four questions' or equivalent wording has been popular. Currently, the paper is divided into two sections: part A consists of a considerable number of short questions, all of which have to be answered, while part B includes longer problems, from which a choice can be made. Here, as elsewhere, the advantage of a larger number of shorter questions is that the influence of chance on the total performance is reduced: it matters less if the candidate starts off on the wrong foot, and makes a mess of one. However, the total exclusion of longer answers would prevent the use of some types of question thought to be valuable.

The paper has not been popular with students, and their performance on it has been consistently worse than on the remaining, more traditional papers. The two facts are probably not unrelated. However, we have been sufficiently convinced of the importance of the general paper as part of the assessment of potential physicists that a good deal of effort has been devoted to altering this reaction. Some years ago, a similar paper was introduced into the examination held at the end of the second year, so that by the time the students take their final examination it is not a novel concept. Problems classes have been organised, with teaching staff in attendance to explain to the students how to set about writing an answer and to give help with carefully chosen examples. Tutors have been provided with material to help and encourage them to devote some of their time with their groups to this kind of activity. Problem sheets, with model answers, have been issued to students to facilitate private study. However, it has not proved easy to change the attitudes and habits of study engendered in the class by some ten years of acquaintance with more traditional examinations. It does seem that this tradition is now changing in schools: if the publication of this book can assist that change, and encourage it to spread into tertiary education, it will have served its purpose.

The Problems

This book contains a selection of the questions that have been used in Bristol over the past 25 years. It is not a perfectly balanced sample: well defined problems are over-represented, but a selection of the more discursive types have been included even though the corresponding answers tend to be disproportionately lengthy. The latter are distinguished by an asterisk placed next to the question, as are other problems which are unsuitable for examination purposes because the solution is too lengthy or too difficult, or requires access to works of reference. Indeed, many of these were not originally used in examinations, but were prepared for use in tutorials, to permit tutors to illustrate their exhortations and advice by reference to a concrete example.

A number of the problems included should be within the competence of a first-year student. No apology is made for this: experience has shown that degree candidates, deeply involved in advanced fields of study, have often forgotten how to handle those basic topics upon which the whole edifice is built and which they may be called upon to use in their work. On the other hand, a few problems on more advanced topics such as atomic spectroscopy or nuclear physics have been included, but these too deal with basic concepts rather than detailed developments. (The students to whom they were presented would not have received instruction in such matters in the previous session so that, to them, they had something of the character of 'general' questions.) In a similar vein, there are a few which would have been standard book-work, or too easy, for a student of, say, engineering or astronomy or even chemistry, but which are, quite properly, 'problems' for a physics student.

The problems are of varying degrees of difficulty, and the answers of varying length. They are not arranged in any kind of order. This is quite deliberate: life is like that. All the questions are original, in the sense that none has been copied directly from elsewhere. However, it is inevitable that those of us responsible for devising them may, on some occasions, have made use—consciously or unconsciously—of an idea which we had seen somewhere else. Many of the originals were phrased in CGS units. Almost all of these have been translated into SI, but one or two have deliberately been left in their original form.

The Solutions

We decided to include the section on answers to make the book more useful to students—not only to permit them to check their own work but also, and perhaps even more importantly, to illustrate the kind of thinking that the problems were designed to stimulate. They do not purport to be 'models', but they are rather more than just notes for guidance. In some cases where part of an answer consists of standard book-work, only a brief outline of the argument has been given. In others, where the argument is more subtle, it has been spelled out in greater detail. But in both cases it will often not be possible to understand what is written without some thought. This is no bad thing: we do not set out to absolve the student from the necessity of thinking. It must be emphasised that where the problem was set as an examination question, the candidate would not have been expected to produce an answer as detailed and as complete as many of those printed here. The scripts would have been marked by an experienced and sympathetic examiner, and a sensible and intelligent attempt at an answer, even though far from complete, would often have been considered quite satisfactory. In a number of instances, attention is drawn to this by enclosing the additional—'bonus'—material in brackets in our answers. A few of the really 'open-ended' questions have given rise to notes of some length: even these do not pretend to be exhaustive treatises, but are intended to indicate how a discussion could ramify in various directions.

On reading through the answers I was struck by the number of times the words 'assume' and 'approximate' occurred. This is just as it should be. The problems call for an element of judgement on the part of the student which is not often present in the more usual type of examination question. In the same way, we do not claim that the answer given is always the only possible answer, or the best answer. We have taken some trouble to ensure that nothing is positively wrong but, allowing for human fallibility, even this cannot be guaranteed. There may be the occasional factor or two out of place: these, like mice, can creep in everywhere. We would of course be grateful to be informed of any errors or suggested improvements, large or small.

We, the authors, are collectively grateful to the University of Bristol for permission to reprint the questions; and I, the editor, am even more grateful to those of my colleagues who have laboured so hard and so long to produce the answers. Any merit which the volume may possess is, without doubt, due to their efforts; the defects are probably my fault.

N Thompson
Bristol, 1987

Part 1

The Problems

1

A cylindrical piece of steel, of diameter 0.4 mm, is lowered very carefully on to water.

Show that it may float horizontally with approximately half its volume above the free surface of the liquid. Assume that the angle of contact is π radians. (Take the density of steel to be $8.5 \times 10^3\ \mathrm{kg\,m^{-3}}$, and the surface tension of water to be $7.0 \times 10^{-2}\ \mathrm{N\,m^{-1}}$.)

2

A planet forms by the gravitational condensation of an initially cold, spherical dust cloud. Find an expression for the radius of the planet if the resulting temperature rise causes the planetary material to reach its melting point. Estimate this radius, using reasonable values of the quantities involved, and comment on the result. Radiation losses may be neglected. ($G = 6.7 \times 10^{-11}\ \mathrm{m^3\,kg^{-1}\,s^{-2}}$.)

3

When told that the world record for the pole vault was about 18 feet, the fast-rising athlete Rod Fibreglass told the press, 'Give me a pole long enough, and I will raise the record to 30 feet'. Could he manage it? How high might he get if he tried hard?

4

A capillary tube of radius r is lowered so that it just penetrates the surface of a liquid of density ρ, surface tension γ and *zero* viscosity, which wets the tube. Describe as fully as you can the motion that ensues.

Liquid helium at temperatures below about 2.2 K approximates to such a liquid, except that there is a critical fluid velocity V_c (with respect to a containing tube) above which the viscosity is much the same as for ordinary liquids. What difference would you expect this to make to the motion?

5

Two soap bubbles, of radii a and b, coalesce to form a bubble of radius c. If the external pressure is P, find an expression for the surface tension S in terms of the quantities P, a, b and c.

If $P = 1$ atmosphere, $S = 30$ dyn cm^{-1} and the radii are of the order of 1 cm, comment on this as a method of determining S.

6

In some recent theories of elementary forces the proton is expected eventually to decay. Derive a lower limit on the proton lifetime from the fact of your own existence, on the assumption that humans will die after exposure to 600 rad (1 rad $= 0.01$ J kg^{-1} energy absorption). Assume that the proton always decays into a positron and a neutral pion and that a neutron bound in a nucleus has the same lifetime and decays into e ($\pm$) and a pion ($\mp$).

Discuss how you could raise this limit as much as possible in a specially designed experiment.

7

The motion of a mass m is controlled by a certain type of spring which produces a restoring force $F = cx^3$ when the mass is displaced by a distance x from its equilibrium position. Discuss qualitatively the form of the free oscillations of the mass m, and show that their period τ is inversely proportional to the amplitude of oscillation x_0. If $c = 10^5$ N m^{-3} and $m = 1$ kg, make a rough estimate of the value of τ when $x_0 = 1$ cm.

8

A large solenoid is required to produce a field of 10 T in a cylindrical volume 5 cm in radius and 1 m long. The space available for the copper windings is 25 cm in radius and 1 m long. Two extreme possibilities for the winding arrangement are:

(i) the 'Swiss roll', with a strip of width 1 m wound around a central tube, which acts as one of the current input connections (the other being at the periphery);

(ii) the 'pancake', with thin discs of outer radius 25 cm, cut along a radius and connected together so that the current flows in a helix around the central region of magnetic field; the current is fed in at the ends of the helix in this arrangement.

Calculate the power needed to produce the required field *using each type of winding*. Which is the more efficient, and why? The resistivity of copper at 300 K is approximately 2×10^{-8} Ω m. You may ignore the space occupied by any insulation, cooling channels etc inside the windings.

9

Calculate how long it would take for water to rise to within 1% of its equilibrium height in a long, vertical capillary tube whose cross section is a rectangle with dimensions 1 mm $\times 10^{-3}$ mm. Assume that the surface tension of water is 7.2×10^{-2} N m^{-1} and that its viscosity is 10^{-3} N s m^{-2}.

10

An air bubble of radius a, rising through a liquid of density σ and viscosity η, has to displace the surrounding liquid as it moves, and because of this it behaves as though it had an inertial mass $M = 2\pi\sigma a^3/3$. If it moves with velocity v, it experiences a retarding force $F = 4\pi\eta av$.

(i) Derive an expression for the vertical distance z moved in time t if the bubble starts from rest at time $t = 0$, neglecting the effect of changing hydrostatic pressure on the size of the bubble.

(ii) Discuss qualitatively how the behaviour will be modified by this effect.

(iii) Again neglecting the complication mentioned in (i), calculate how long a bubble 1 mm in diameter, starting from rest, takes to rise (*a*) 5 cm and (*b*) 10 cm through water. ($\eta = 10^{-2}$ g cm^{-1} s^{-1}.)

11

A straight tunnel is bored through the earth to connect two points A and B on the surface. Show that under certain assumptions, which you should state, the time T to fall freely from A to B is independent of A and B. Calculate T in minutes, taking the earth's radius to be 6400 km.

Comment on the feasibility of a web of such tunnels as a global transportation system.

Does a straight tunnel provide the quickest connection between A and B? Discuss briefly.

12

One method of measuring the surface tension γ of a liquid is to measure the mass m of the drops which drip from the bottom of a tube of radius r, when the liquid is fed slowly in at the top. Elementary theory suggests that $mg = 2\pi r\gamma$. Experiment shows that

$mg = 2\pi r\gamma f(a/r)$, where the function f depends on the ratio of r to a 'characteristic length' a of the liquid. Use dimensional arguments to find the form of a, and discuss qualitatively the form you would expect for f.

The experiment gives incorrect answers if the drops form too rapidly. Discuss the physical phenomena which influence the maximum permissible rate of formation of drops.

13

Sound cannot propagate through a vacuum. In a well known demonstration of this, an electric bell is hung up inside a bell-jar, which is then evacuated. Sure enough, as the air pressure is reduced the noise of the bell becomes fainter and fainter; the bell becomes quite inaudible when the pressure falls below about 1 cm Hg. Explain in detail why this demonstration is quite spurious, and what the true interpretation is.

14

---

*Given that oil is available as a raw material for domestic heating, consider the following options.

(i) The oil is burned in a furnace that heats the house directly.

(ii) The oil is burned in a power station to generate electricity, which is used to heat the house by means of electric fires.

(iii) The oil is burned in a power station to generate electricity, which is used to heat the house by means of a heat pump.

Discuss the overall efficiency in each case and compare them.

15

You are invited to develop a theory of bottle washing. Assume that you have a large volume W of clean water and a bottle of volume B, which contains a small volume D of dirt ($W \gg B \gg D$).

When water is put into the bottle the dirt dissolves immediately, and when the bottle is 'emptied' a small residue R of solution remains. How do you get the bottle as clean as possible using all the water?

16

A small specimen of material of magnetic susceptibility χ is placed on the axis of a flat circular coil, at some distance from the plane of the coil, and is made to oscillate back and forth along the axis with a small amplitude. A uniform magnetic field parallel to the axis is established over the whole system. Derive an expression for the EMF developed in the coil, assuming that the coil radius is very much less than the coil–specimen distance.

17

A suggested route to controlled nuclear fusion involves heating plasmas with beams of relativistic electrons. According to a recent issue of *New Scientist,* 'so high is the electric current supported by beams of this kind that they are self-focusing, being 'pinched' by their own magnetic field'. Discuss this statement.

18

(*a*) A film of a horse race, taken using a telephoto lens, shows the horses strangely foreshortened as they gallop towards the camera.

(*b*) A photograph of skyscrapers taken with a wide-angle lens shows them distorted into barrel-like columns.

Explain these observations.

19

Two stars in a binary system have a separation $2r$ and equal masses m, and move in circular orbits about their centre of mass. One star explodes by expelling a small fraction of its mass very rapidly, and immediately after this its recoil speed is v_f, What is the largest value of v_f for which the two stars will remain gravitationally bound?

20

Explain (qualitatively and quantitatively) how the source of the Mississippi river can be about 5 km closer to the centre of the earth than is its mouth. (The earth's radius is 6400 km, the source of the Mississippi is at latitude 50°N and the mouth is at latitude 30°N.)

21

A glass slide is coated on one side by successive evaporation of a thin half-reflecting layer of silver, a layer of silica 4000 Å thick with a refractive index $\mu = 1.5$, and a second half-reflecting layer of silver. It is then painted black, leaving clear only a long central strip 60 μm wide. The slide is illuminated with a beam of parallel white light at normal incidence, and the light passing through is collected by a large convex lens whose focal length is 20 cm. Describe in a semi-quantitative manner the variation of illumination in the focal plane of the lens, assuming a uniform spectral distribution of the intensity of the incident light.

22

In the ground state of atomic hydrogen, the wavefunction of the electron is finite (though small) at distances from the nucleus such that the potential energy is greater than the total energy. Comment on the probability of observing the electron at such a distance from the nucleus.

23

An 'electrostatic precipitator' could take the form of a rod at a high potential along the axis of a cylinder. Waste gases containing dielectric particles (initially uncharged) flow along the cylinder. Derive an expression connecting the applied potential, the dimensions of the equipment, the mean flow velocity and any other relevant parameters, which could be used in the design of an installation to remove the dust.

24

It is well known that in an orbiting space vehicle the occupants can drift around freely in 'zero-gravity' conditions. Assume that you are in a (strong) spaceship, 100 m long and fairly narrow, which is in a circular orbit of 1000 km radius around a neutron star, with its axis always pointing towards the centre of the star. There is an inspection tunnel running down the centre of the ship. What would happen if you attempted to float down it? Calculate likely values of the acceleration observed, assuming that the mass of the star is 3×10^5 earth masses and that the radius of the earth is 6×10^6 m.

Is an orbit such that the axis of the craft always points toward the star a stable condition? Explain briefly.

25

---

*A system using a tethered balloon has been proposed as a means of providing television coverage for an area 1000 km across. How high would the balloon have to be to be able to transmit to the whole area by line of sight?

Discuss the suitability of steel or kevlar (a high-strength polymer) for the tethering cable. Identify some of the main problems that would be encountered in making such a system work. (The tensile breaking strength of steel is $1.6 \times 10^9 \text{ N m}^{-2}$, that of kevlar is $2.0 \times 10^9 \text{ N m}^{-2}$.

26

A conducting sphere of radius a and mass m, hanging by an electrically conducting wire of length l from an earthed support, is in equilibrium midway between two vertical parallel earthed metal plates a distance $2b$ apart.

The plates are then given potentials $+V$ and $-V$ with respect to earth. Show that, if V is large enough, the sphere will be attracted to one or other of the plates, and calculate approximately the least value of V for which this happens. (Assume $b \gg a$.)

27

A 1 mW laser beam of 2 mm diameter and wavelength 600 nm is focused by a lens of 5 mm focal length onto the surface of an opaque material.

(i) Neglecting thermal conduction, show that temperatures similar to that of the surface of the sun could be achieved.

(ii) Including thermal conductivity, estimate roughly what temperature would be reached if the material were lead. Would any lead melt?

Stefan's constant is $\sigma = 5.7 \times 10^{-8} \text{ W m}^{-2} \text{K}^{-4}$. The thermal conductivity of lead is $30 \text{ W m}^{-1} \text{K}^{-1}$.

28

A loop of steel wire, carrying a weight of 10 kg, hangs over a block of ice at 0°C. Estimate the rate at which the wire sinks through the ice. To simplify the calculations, assume that the portion of wire inside the block at any time can be treated as an infinitely rigid horizontal bar of length 0.2 m and 1 mm square cross section. (The density of ice is $9.17 \times 10^2 \text{ kg m}^{-3}$; the latent heat of fusion of ice is $3.36 \times 10^5 \text{ J kg}^{-1}$; the thermal conductivity of steel is $63 \text{ J m}^{-1} \text{s}^{-1} \text{K}^{-1}$; the Clausius–Clapeyron equation is $dp/dT = L/T(v_2 - v_1)$, where p = pressure, T = temperature, L = latent heat and v_1 and v_2 are specific volumes.)

Comment on any assumptions and approximations made.

29

A superconductor is effectively a perfect diamagnetic material, so that $B = 0$ everywhere inside it. A small bar magnet with a weight of 3 g and a magnetic moment of 0.1 $\mathrm{A\,m^2}$ is lowered towards a horizontal superconducting plane. At some height above the plane, the magnet floats freely. Estimate this height.

30

An isolated conducting sphere of radius 10 cm is bombarded by a broad beam of 1000 eV electrons, with a uniform current density.

Assume that all the electrons that strike the sphere are captured and that no charge is lost. When a steady state is reached:

What is the potential of the sphere?
What is the charge on the sphere?
What is the field at the surface of the sphere?

Discuss the assumption that no charge is lost.

31

*Tritium decays with a half-life of 12 years and emits a beta particle whose range in most materials is of the order of a few microns. In a photographic emulsion, a grain of silver halide containing a beta track is reducible to silver by normal photographic development. It is proposed to use this effect to obtain a 'contact print' showing the distribution of water in a frozen, sectioned sample of a clay previously doped with tritiated water. Estimate the dosage necessary to get a good print with resolution 10^{-1} mm with an exposure time of one hour. What practical difficulties might arise?

32

In a gas-filled electron multiplier, two regions of low electric field are separated by an electrode constructed from parallel wires whose diameters are very much smaller than their separation. The electric fields in the two regions are normal to the plane of the electrode and are in the same direction, but may be of different magnitudes. Free electrons exist in the two regions.

Estimate the fraction of the electrons drifting towards the electrode screen that will penetrate it, in terms of the fields on either side. Would the situation be changed if the electric fields were very high? Would the result be different if the device were evacuated?

33

The uncertainty principle prevents a pencil from balancing on its tip for longer than about four seconds. Confirm this impertinent intrusion of quantum mechanics into matters mundane. (Hint: For a pencil of length l the displacement varies with time as $\theta(t) = \theta_0 \cosh(t/\tau) + \dot{\theta}_0\tau \sinh(t/\tau)$, where θ_0 and $\dot{\theta}_0$ are the initial small angular displacement and angular velocity away from the vertical, and $\tau = (2l/3g)^{1/2}$.)($\hbar = 10^{-34}$ J s.)

34

A simple pendulum with mass m has a length l which is slowly decreased by pulling the string up through the point of suspension. Show, in the limit of small oscillations and slow changes in l, that the energy of oscillation E is proportional to the frequency f.

35

Huxley is alleged to have said that 'six monkeys, set to strum unintelligently on typewriters for millions of years, would be bound in time to write all the books in the British Museum'. Kittel (*Thermal Physics,* 2nd edn, p 53) shows that this is nonsense.

Could any array of computers acting at random within the time and distance scales of our universe produce even a single book? We can tolerate a proportion of errors (we frequently do): will this make any difference? Would it make any difference if we were to be satisfied with any book whatsoever?

36

The 'gun' of a CRT consists of a small hot filament F at $x = 0$ and an aperture at $x = d$. Between $x = 0$ and $x = d$ a uniform electric field E accelerates the electrons to velocities which are non-relativistic but are considerably larger than their initial thermal velocities. Show that when they arrive at the plane $x = d$ they appear to have travelled in straight lines (i.e. in field-free space) from an 'image' source I at $x = -d$.

A uniform magnetic field B is now applied axially. Is it still valid to regard I as an image of F?

37

The threshold momentum at which an energetic proton may collide with a stationary free proton to produce a nucleon–antinucleon pair can be calculated to be 6.5 GeV/*c*. Experimentally, the threshold momentum using a copper target was found to be only about 4.8 GeV/*c*. How do you account for this result? State carefully and precisely how momentum and energy are conserved in the collision. How do you expect the reaction probability to fall off well below the 'ideal' threshold of 6.5 GeV/*c*?

38

A rocket is descending to earth with a resultant velocity greater than that of sound, although its vertical component is less than that of sound. The effect of the acceleration due to gravity and of the air resistance may be neglected. Calculate the time of arrival at a point on the ground of a sound wave from a point on the trajectory. Hence show that some points on the ground will experience a shock wave. Show that the loci of points on the ground at which the shock wave arrives at a constant time are ellipses.

39

It is proposed to make a man-powered helicopter with a rotor 10 m in diameter. Assuming that the rotor blows a cylindrical column of air uniformly downwards, the cylinder diameter being the same as the rotor diameter, and that the weight of man plus machine is 200 kg, calculate the minimum mechanical power (in watts) that it is necessary for the man to generate if he is to remain airborne. Is the system practicable? The density of air is 1.23 $\mathrm{kg\,m^{-3}}$.

40

---

*A small particle of graphite, roughly spherical and about 10 μm in diameter, falls in vacuum through a horizontal beam of 3 keV protons. It has been shaken loose from a surface about 4 cm above the top edge of the beam. The proton beam carries a current of 10 mA spread uniformly over a circular cross section 20 mm in radius. You may assume that a 3 keV proton cannot penetrate 10 μm of graphite, so that any proton that hits the graphite stays in it. How much can you predict about what will happen to the graphite particle? (C_p for graphite is 800 $\mathrm{J\,kg^{-1}\,K^{-1}}$.)

41

A pendulum is made by hanging a heavy steel bar from a horizontal axle which passes through a hole near one end of the bar. The sliding surfaces at the pivot are lubricated in such a way that the friction between them either (*a*) increases or (*b*) decreases as the sliding velocity increases. The axle is now rotated rapidly at a constant angular velocity and, because of the friction, the equilibrium position of the bar is slightly deflected from the vertical. Examine small oscillations of the pendulum about the equilibrium position under the two conditions (*a*) and (*b*).

42

Show that the conservation of energy and momentum require that if a collison between a fast electron and an electron at rest is to lead to the creation of an electron–positron pair, the kinetic energy of the fast electron must be at least $6mc^2$, where m is the mass of the electron.

43

---

*A sheet of polyethylene (of relative dielectric constant 2.3) is 5 mm thick, and contains a small spherical cavity filled with air at STP. If the sheet is used as the dielectric in a parallel-plate capacitor, what is the maximum steady voltage that can be applied across the capacitor before the air in the cavity breaks down, and what will happen when it does break down? (Take the breakdown field of air to be 100 $\mathrm{kV\,cm^{-1}}$.)

How will the breakdown behaviour be modified if the polyethylene, instead of being a perfect insulator, has a very small but finite conductivity, say $10^{-15}\,\Omega^{-1}\,\mathrm{cm}^{-1}$?

44

Show that a wheel rolling freely on a flat horizontal rough surface can never be slowed down by static friction. Experimentally, such wheels do slow down, at a rate corresponding to an effective 'coefficient of friction' of less than 0.001, and microscopic examination fails to reveal any slipping. Suggest a possible mechanism for the slowing down.

45

According to the *Observer* Colour Magazine a *demi-cannon* recovered from the Tudor ship *Mary Rose* weighed 5 cwt and fired a 32 lb shot. Comment on these figures, assuming that the *demi-cannon* had an extreme range of about one mile. (For the purposes of this problem, you may take the weights to be 300 kg and 15 kg respectively, and the range to be 1.6 km.)

46

A total charge $-q$ is spread uniformly on the surface of a thin insulating spherical shell of radius a. Thus there is no internal field. A point charge $+q$ is placed at the centre of the sphere. There is now no external field. The point charge can be displaced a distance $r < a$ from the centre without the expenditure of any work. However, this gives the system a dipole moment rq with a corresponding external field. How do you explain the apparent paradox of the appearance of this external field energy?

47

---

*In order to ensure safe re-entry into the atmosphere, manned space vehicles use a heat shield that is able to ablate (evaporate). Discuss some of the mechanisms by which the kinetic energy of the vehicle is dissipated on re-entry, and the factors which determine the rate of ablation.

48

Show that when a plane soap film is punctured, the speed at which the radius of the hole increases is $(2T/\sigma)^{1/2}$, where T is the surface tension of the soap solution and σ is the mass per unit area of the film. How fast is this for a typical bubble?

49

A billiard ball of radius a rests on a table. It is hit with a cue in such a way that it starts out with speed u_0 and backspin ω_0 about a horizontal axis perpendicular to the direction of motion. How does the subsequent motion depend on the ratio $u_0/a\omega_0$?

50

A fuse consists of a short length of fine copper wire, mounted in a glass tube between end contacts which can be assumed to stay at room temperature. What determines the current I_c at which the fuse will melt? If a current slightly greater than I_c is suddenly applied, what determines the time the fuse will take to melt?

51

A simple pendulum consists of a spherical bob hanging from a thread 1 m long. The upper end of the thread is held between finger and thumb, and the hand is moved to and fro slightly in a horizontal plane at the natural frequency of the pendulum, to keep it swinging. It is found that the amplitude of swing of the bob is 0.1 m when the amplitude of motion of the hand is 2.5×10^{-4} m. If the hand is now held still, how long will it take for the amplitude of the bob to decay to 10/e cm? What else would you need to know about the pendulum in order to estimate the viscosity of air, and how reliable do you think this estimate would be?

52

Suppose that you are a consultant in mechanics to the British Olympic Skating Team, and that you wish to improve their aerial spins. Devise a simple model for such spins. What aspects of technique would you advise the team to concentrate on? What is the *maximum* number of turns you would expect to be possible between take-off and landing?

53

*In an experiment to test the self-heating of a platinum resistance thermometer the following values of i and V were transcribed from a laboratory notebook.

i(mA)	V(μV)
2	2201
3	3302
4	4468
5	5514
6	6624
7	7736
8	8859
9	9982
10	11109

The uncertainties in V (due to the finite sensitivity of the galvanometer used in the potentiometer circuit) are expected to be about 1 μV; the values of i may be assumed to be exact.

(i) Are there any oddities in the data, and if so, what should you do about them? Can you explain them?

(ii) Is there any significant self-heating, and if so, how would you describe it?

(iii) What is the best value of the thermometer resistance for the purposes of specifying temperature?

(iv) What measuring current would you recommend be used, and why?

54

It is sometimes stated that it would be possible to construct a spaceship using a photon drive, which would be able to travel away from the earth with a speed close to that of light. Assuming that the fuel consists of equal masses of protons and antiprotons, does one get a greater final velocity (*a*) by allowing half the fuel to annihilate and using the energy released to eject the remaining half or (*b*) by allowing all the fuel to annihilate and ejecting photons?

55

---

*A chemical balance is used to weigh an hourglass. Initially, all the sand is in the upper half of the hourglass, and the passage to the lower half is temporarily blocked. At time $t = 0$ the blockage clears, and thereafter sand falls continuously into the lower half of the glass until the upper half has been emptied.

Discuss, as quantitatively as you can, what variation (if any) is to be expected in the indicated weight of the hourglass during this process. (The balance is capable of detecting changes in mass of 1 μg, and the hourglass contains 1 kg of sand. The time taken for the transfer of the sand is, of course, 1 hour.)

56

A steel wire is stretched horizontally between two supports 100 m apart. If the tension in the wire is half the breaking tension, how far does the middle of the wire sag below the true horizontal?

If the system is set up on a hot summer's day, will the wire break during a cold winter? (The density of steel is 8×10^3 kg m^{-3}; the tensile strength is 2×10^9 N m^{-2}; Young's modulus is 2×10^{11} N m^{-2}; the thermal expansion coefficient is 10^{-5} K^{-1}.)

57

For practical purposes rubber is incompressible, and when a sheet of it is stretched with equal forces acting in two directions at right angles within the sheet we find that

$$\sigma = \frac{E}{3}(\lambda^2 - 1/\lambda^4)$$

where σ is the true stress, E is Young's modulus and λ is the extension ratio l/l_0, l being the stretched length of an element whose unstretched length was l_0. This relation is valid for $1 < \lambda < 2$; for larger λ, σ increases more rapidly.

When a spherical rubber balloon is only slightly inflated it has a radius r_0 and a wall thickness t_0. Investigate the relationship between the radius r of the inflated balloon and the internal pressure excess p, and comment on your result.

58

A homogeneous mixture of uranium and carbon forms the core of a simple nuclear reactor. Suppose that

(i) the core is a cube of side L;

(ii) the neutrons emitted by the uranium travel with an average velocity v and a mean free path λ;

(iii) neutrons are captured, on average, after N collisions;

(iv) for each neutron captured, k new neutrons are emitted.

Show that the rate of change of neutron density n in the core is given by

$$\frac{\mathrm{d}n}{\mathrm{d}t} = nv\left(\frac{k-1}{\lambda N} - \frac{3}{2L}\right).$$

Estimate the size of the pile required to achieve a chain reaction, given that $\lambda \simeq 10$ cm, $k = 1.04$ and $\mathrm{N} \simeq 100$.

59

It is observed that the acceleration g due to gravity is greater down a mine than on the earth's surface. Show that this result can be explained if the earth's density increases sufficiently rapidly with depth (derive a precise condition).

60

You have no mackintosh or umbrella, and have to make a journey on foot in steadily falling rain. If you run, the journey will not take so long, but you may intercept more rain. Taking as your criterion the necessity to minimise the number of rain drops that strike you, and assuming that the rain falls steadily and vertically at $10\ \mathrm{m\,s^{-1}}$, construct a theory that enables you to decide the best speed at which to run. Mention any shortcomings of the theory which occur to you.

61

A 50p piece may be taken as a heptagonal disc of uniform material, except that the sides, instead of being straight, are arcs of circles of radius 3 cm centred on the opposite vertex.

Calculate approximately how fast this coin can roll along a horizontal plane without losing contact, assuming that as long as there is contact there is no slipping.

62

A snow surface has low cloud overhead. Rain falls. Estimate the rate of precipitation that would transfer the same amount of heat to the surface as it receives from radiation. Assume that the snow and cloud base act as black bodies, that the rain is at the same temperature as the cloud base, and that the snow is at the melting point. Stefan's constant is $5.67 \times 10^{-8}\ \mathrm{W\,m^{-2}K^{-4}}$. How would your result be affected if the cloud were too thin to act as a black body?

63

In the CERN Intersecting Storage Rings, beams of protons of kinetic energy 24 GeV (i.e. with a velocity approaching c) carry currents of up to 30 A in two rings of circumference 950 m.

Calculate the total charge carried by the protons in one of these rings, and the total energy deposited when the beam is ejected from the ring and dumped in a copper block. To what temperature will this block be raised if its mass is 100 kg? (The specific heat of copper is $3.8 \times 10^{2}\ \mathrm{J\,kg^{-1}K^{-1}}$.)

64

A small moon of mass m and radius a orbits a planet of mass M while keeping the same face towards the planet. Show that if the moon approaches the planet closer then $r_c = a(3M/m)^{1/3}$, loose rocks lying on the surface of the moon will be lifted off.

65

A circular coil of copper wire with a total length of 500 m and a resistance of 40 Ω, rotating about its axis so that its speed is $55.2\ \mathrm{m\,s^{-1}}$, is suddenly brought to rest. The ends of the coil are connected to a ballistic galvanometer which shows a deflection at the instant that the coil is stopped; the deflection corresponds to a total charge of 3.5×10^{-9} C. Calculate a value of e/m for the conduction electrons in copper.

66

A flat plate is inclined at an angle α to the horizontal and is covered by a thin film of water, which wets the plate and flows under the combined effects of gravity and viscosity. Prove that the volume q of water which crosses unit horizontal length in unit time is related to the thickness d of the film by

$$q = \rho g \sin\alpha\, d^3/3\eta$$

where ρ and η are respectively the density and viscosity of water and g is the acceleration due to gravity.

Neglecting surface tension and the loss of water by evaporation, show that the equation that determines the change of d with time is

$$\eta \frac{\partial d}{\partial t} + \rho g \sin\alpha\, d^2 \frac{\partial d}{\partial x} = 0$$

where x is the distance along the plate in the direction down the slope.

Verify that the general solution of this equation is

$$\eta x - \rho g \sin\alpha\, d^2 t = F(d)$$

where F is an arbitrary function.

Assuming that, initially, the film thickness is proportional to the distance from the top of a plate of length l, and that water drips off at the bottom, show that for sufficiently long times the thickness near the bottom is $(\eta l/\rho g t \sin\alpha)^{1/2}$.

If $l = 10$ cm, $\alpha = 5°$ and $\eta = 1.3 \times 10^{-3}$ N s m^{-2}, calculate this thickness after one hour.

67

Explain how the plates of the capacitor in an *LC* circuit should be vibrated mechanically in order to cause the amplitude of alternating current in the circuit to increase.

68

A nuclear reaction of the type a + A → b + B is induced by bombarding unpolarised nuclei A with a beam of unpolarised projectiles a. Use symmetry principles to show that if the products b and B are polarised, they can only be polarised perpendicular to the plane of scattering. Show also that if the particles going to the left at an angle θ are polarised 'up', then those going to the right at an angle θ must have an equal polarisation 'down'.

69

Does any entropy change occur when a shuffled pack of cards is sorted into a regular arrangement? If so, can any useful cooling be achieved by carrying this out adiabatically?

70

In certain laser experiments on ionised gases the free electrons are in an oscillatory electric field which is constant in direction, but whose amplitude varies slowly along the direction of the field, while remaining constant at right angles to this.

Show that over times long compared with the time period of the field, the free electrons will drift to places where the field is a minimum.

71

---

*A programmer spends most of the working day at a desk in a large (10 × 10 × 3 m) but not specially ventilated room. In the same room, experiments involving benzene and mercury are in progress but the programmer does not come into direct contact with them.

After a couple of years the programmer complains of mysterious health problems, and eventually sues the company for negligence. The company denies liability, and claims that at no time is more than a few grams of either liquid exposed to the air.

As a consultant physicist, what advice would you offer to either party? (Extract from *Dangerous Properties of Industrial Materials*: 'Thresholds below which no serious health hazard is likely are, for Hg: 0.05 mg m^{-3} and, for benzene: 30 mg m^{-3}.) You may wish to make use of the following quantities.

Vapour pressure of mercury at room temperature $\simeq 2 \times 10^{-7}$ atm.
Vapour pressure of benzene at room temperature $\simeq 0.13$ atm.
1 atm $\simeq 10^5$ N m^{-2}.
Molecular weight of mercury $\simeq 200$.
Molecular weight of benzene $\simeq 80$.
Gas constant $R \simeq 8$ J K^{-1} mol^{-1}.
Avogadro's number $Z \simeq 6 \times 10^{23}$ mol^{-1}.

72

A Dewar containing liquid helium at 4.2 K has above it a static column of gas 50 cm long and 1 cm in diameter, the top of which is at room temperature. Estimate the boil-off rate of liquid helium due to thermal conduction down the column. The latent heat of helium is 2.75 J cm^{-3}; m(proton) $= 1.67 \times 10^{-27}$ kg; $k_B = 1.38 \times 10^{-23}$ J K^{-1}.

73

The number of beta rays emitted per second by fission fragments produced in an explosion is found empirically to take the form

$$\mathrm{d}N = kT^{-1.2}\,\mathrm{d}T$$

where T, the time, is measured from the time of the explosion, and k is a constant. Such a formula is accurate to within a factor of about 2 over the wide range of time 1 s $< T \leqslant 100$ years.

In order to show that such behaviour is not unexpected, we may start from the fact that the lifetime τ of a beta-active nucleus is given approximately by the formula $\tau \propto E^{-5}$ where E is the disintegration energy. By considering the distribution of beta ray disintegration energies that are to be expected among the decaying fission fragments, and making appropriate simplifying assumptions, show that a distribution of lifetimes $\mathrm{d}N \propto \tau^{-1.2}\,\mathrm{d}\tau$ results.

Show further that this distribution of lifetimes leads to the result quoted for the variation of activity.

74

A small permanent-magnet DC electric motor, as used in model trains, may have its speed varied by inserting a series resistor to decrease the current or by switching the current with a transistor so that the full voltage is applied for a fraction of the time. It has been claimed that the latter method gives less resistive power loss.

To determine whether this claim is valid, consider the case where the speed is reduced by a factor of two by either method. Assume that the speed is proportional to the mean torque of the motor and does not vary significantly during the pulse period. Assume also that the resistive loss across the transistor is negligible and that the resistance of the motor is such that the current drawn by the motor at full speed is 0.1 times the current drawn when it is held stationary with full voltage applied.

Calculate the power dissipated in the motor for each method. How do these values compare with the dissipation at full speed?

75

A plausible design for a perpetual motion machine is as follows. A wheel of radius r is free to turn on a horizontal axle, and has four point charges $+q$ spaced equally around its perimeter. It is half immersed in oil of dielectric constant ϵ (so that the axle lies in the surface of the oil), and a fixed point charge $-Q$ is placed on the surface of the oil, on one side of the wheel and in the same plane, at a distance $R > r$ from the axle.

Starting with two of the charges q vertically above and below the axle, one then expects the attraction exerted by Q on the upper charge to exceed that on the lower charge by a factor ϵ, so that the wheel should start to rotate, and continue to do so indefinitely. Unfortunately it does not work. Explain why not. What position would you expect the wheel to take up?

76

A thin-walled cylindrical tube of circular cross section, filled with air, has ends which are spanned by soap films. The soap films are not necessarily flat. Discuss the equilibrium and stability of the system for various amounts of air in the tube. (You may use the following formulae if you wish, but you are recommended to try to answer the question without them. The volume of a spherical cap of radius R and height k is $\pi k^2(3R - k)/3$; the surface area is $2\pi Rk$.)

77

A dust particle in an ionised gas acquires an electric charge, even when the gas is electrically neutral overall, because the numbers of electrons and ions striking the particle each second are not equal. Consider a spherical dust particle of radius r which is placed in ionised hydrogen gas with a temperature T and a density of n protons and n electrons per unit volume. Derive expressions for

(*a*) the initial rate of accumulation of charge on the particle,

(*b*) the maximum potential reached at the surface of the particle, and

(*c*) the time scale for this maximum potential to be reached.

78

Would it be necessary to change the focus of a telescope to look at the moon after focusing it on a star? How big would the telescope have to be for any refocusing to be needed? The moon's distance from earth is 3.8×10^8 m.

79

You have a cup of hot coffee and add cold milk. You find it is still too hot to drink. If you wish to drink that coffee quickly would it have been quicker to wait before adding the milk? Explain your answer.

80

Students in a laboratory class are asked to investigate the rate at which ether contained in a vertical capillary tube evaporates, as measured by the rate of fall of the liquid level.

Work out a simple theory for this experiment, assuming the changes to be sufficiently slow that it is a good approximation to say that the situation is one of dynamic equilibrium at all times. Consider what other assumptions are necessary. Making these as simple as possible, deduce a relation between time elapsed and length of column evaporated.

81

A recent natural history film showed the so-called camphor fish. This tiny fish (about 3 mm in diameter) spends much of its time floating at the surface of the water. It can, in moments of stress, exude a camphor-like substance from its tail, as a result of which it shoots rapidly across the water. Give a theoretical model for this effect and estimate to an order of magnitude the maximum velocity that might be reached. (The surface tension of water is 72 $\mathrm{mJ\,m^{-2}}$; the viscosity of water is $10^{-3}\ \mathrm{kg\,m^{-1}\,s^{-1}}$.)

82

A clock in a satellite appears to an observer on earth to be running at a rate governed by two factors.

(i) Ordinary relativistic time-dilation.

(ii) The effect of its acceleration in the gravitational field of the earth, which is to increase the clock rate by a fraction $\Delta\phi/c^2$, where $\Delta\phi$ is the difference in gravitational potential between the positions of the clock and the observer.

Discuss the combined effect of these two factors on the observed clock rate, as a function of the radius of the orbit of the satellite. Would you expect the effect to be (*a*) measurable, (*b*) important?

83

A well known apparatus for demonstrating electromagnetic induction consists of a vertical iron (or laminated iron) post about 50 cm high and 5 cm in diameter, with a coil of wire wound around its lower end. If a loose-fitting metal ring is lowered over the post after an AC supply to the coil has been switched on, it will float at a certain height above the coil. Explain why, and derive an expression for the upward force on the ring. What determines the height at which it floats?

84

Two dipoles (e.g. permanent magnets) of moments M_A and M_B are situated at points A and B, which are a distance x apart. The axis of M_A is perpendicular to the line AB, while that of M_B lies along AB. Calculate the force and/or couple which each exerts on the other, and comment on the results.

If two equal magnets are floated on the surface of a liquid, arranged as above, and then released, discuss qualitatively their subsequent motion and final position, neglecting any effects due to the magnetic field of the earth, and assuming the liquid to be sufficiently viscous to prevent oscillations.

85

A compound pendulum consists of a thin cylindrical rod of length $l_1 + l_2$ and uniform cross section; the pivot is at a distance l_1 from the top. The material above the pivot is of density ρ_1 and has a coefficient of linear thermal expansion α_1, the corresponding quantities for the material below the pivot being ρ_2 and α_2. Consider whether it is possible to make the period independent of temperature.

86

A man of mass M standing on a heavily damped weighing machine throws a ball of mass m upwards, catches it as it falls, and keeps on throwing it up and catching it. Assuming that the weighing machine gives a true reading of the time-averaged force on it, calculate the value of this reading. Explain your reasoning carefully, so that the logical structure of the argument is as clear as you can make it.

87

Two 'standard metre' bars, placed end to end, will both contract a little, owing to their mutual gravitational attraction. (*a*) Give an order-of-magnitude calculation of the amount of contraction. (*b*) Calculate it more exactly, assuming the bars to be long compared with their transverse dimensions.

88

In the Millikan experiment to determine the electronic charge e one measures the charges $q_1, q_2, \ldots, q_n$ on n oil drops and calculates e as the lowest common factor of these charges. However, in any particular experiment bad luck might result in all the oil drops having multiples of two, or more, charges on them, so that the calculated e might be several times larger than the true value. What is the chance of this?

89

A uniform vertical pipe of length h and radius a is closed at the bottom, filled with water, and connected at the top to a shallow tank of water of very large capacity. The bottom of the pipe is opened at time $t = 0$. Neglecting viscosity, and neglecting the kinetic energy of the water in the tank, derive an expression for the velocity of the water in the pipe at time t, and show that after a sufficiently long time it tends to $(2gh)^{1/2}$.

Show that if viscosity is not neglected, the velocity at the centre of the tube tends after a sufficient time to the limiting value $v_0 = (4gh + k^2)^{1/2} - k$ where $k = 8\eta h/\rho a^2$.

90

Imagine that mass, which governs the acceleration of bodies and also their mutual gravitational interaction, might sometimes be negative. Consider two masses M_1 and M_2, isolated in space and initially at rest a little way apart. What happens if

(*a*) $M_1 > 0$, $M_2 > 0$;
(*b*) $M_1 < 0$, $M_2 < 0$;
(*c*) $M_1 > 0$, $M_2 < 0$, $|M_2| < M_1$;
(*d*) $M_1 > 0$, $M_2 < 0$, $|M_2| > M_1$?

91

In a betatron, particles are accelerated on circular orbits by a rotational electric field produced by a varying magnetic flux. Show that if the magnetic flux through an orbit of radius r_0 at any instant is $2\pi r_0^2 B$, where B is the field at the orbit, then the radius of the orbit remains constant. Show that the orbits will be stable if the magnetic field in the region $r \simeq r_0$ varies as $1/r^n$ with $n < 1$.

92

An earth satellite, supposed to be in a geostationary orbit, in fact drifts from west to east at the rate of one complete revolution per month. Calculate the change of orbital speed needed to bring it into the correct orbit. In which direction should the force be applied? (The earth's radius is 6400 km.)

93

A ship has a period of roll T. It steams at speed V through ocean waves which have a period τ and speed v. Find the condition for which there are four 'sensitive courses', where the ship encounters waves coinciding with its rolling period.

94

The following problem appears to have been first mentioned by Lewis Carroll.

A long light cord passes over a light frictionless pulley. At one end of the cord a bunch of bananas is tied, and at the other, there is a monkey whose mass is equal to that of the bananas. If the system starts at rest, with the bananas higher than the monkey, what will happen if the monkey climbs the cord?

You are invited to show that provided each side of the cord remains vertical, the bananas ascend at the same rate as the monkey. You are also invited to generalise the problem in two separate ways.

(i) Assume that the pulley is still frictionless but has a finite moment of inertia.

(ii) Neglect the friction and moment of inertia of the pulley, but consider the possibility that the monkey's movements may cause him to swing like a pendulum as he climbs.

95

My grandfather clock was recently overhauled. On return I found that it was losing four minutes a day and the screw which should adjust the length of the pendulum was too stiff to move. I decided to regulate it by fixing a small weight to the rod.

(i) Why does this work?

(ii) At what point on the rod should a weight be put to have the maximum effect?

(iii) If the weight is put at this point how big should it be if the pendulum bob weighs 1 kg?

96

This question concerns the behaviour of the following quantities: displacement ($\boldsymbol{r}$), time (t), velocity ($\boldsymbol{v}$), momentum ($\boldsymbol{p}$), spin ($\boldsymbol{s}$), electric field ($\boldsymbol{E}$) and magnetic field ($\boldsymbol{B}$) under transformations of space inversion (P) and time reversal (T).

(i) Which of these quantities change sign under P?

(ii) Which change sign under T?

(iii) Use your answers to (i) and (ii) to decide whether each of the following laws (*a*)–(*d*) is invariant under P, under T, under both or under neither. Show your reasoning clearly.

(*a*) The Lorentz force law for a particle of charge q:

$$\frac{\mathrm{d}\boldsymbol{p}}{\mathrm{d}t} = q(\boldsymbol{E} + \boldsymbol{v} \times \boldsymbol{B}).$$

(*b*) Newton's laws of motion for two mutually gravitating bodies with masses m_1, m_2:

$$m_1 \frac{\mathrm{d}^2 \boldsymbol{r}_1}{\mathrm{d}t^2} = -\frac{Gm_1m_2(\boldsymbol{r}_1 - \boldsymbol{r}_2)}{|\boldsymbol{r}_1 - \boldsymbol{r}_2|^3} = -m_2 \frac{\mathrm{d}^2 \boldsymbol{r}_2}{\mathrm{d}t^2}.$$

(*c*) A law that predicts that μ^- particles with spin $\boldsymbol{s}$ will decay to produce electrons whose velocity $\boldsymbol{v}$ is related to $\boldsymbol{s}$, on average, by

$$\langle \boldsymbol{s} \cdot \boldsymbol{v} \rangle = -hc/36\pi.$$

(*d*) A law that predicts that a neutron with spin $\boldsymbol{s}$ in a magnetic field $\boldsymbol{B}$ has potential energy

$$u = \alpha \boldsymbol{s} \cdot \boldsymbol{B}.$$

97

As is well known, absolute zero cannot be reached, according to macroscopic thermodynamics. However, there is a finite probability that the lattice of a finite body will be free from phonons. Suppose you had an evacuated cavity that could be cooled to 10^{-5} K and could suspend a particle within it without dissipation; roughly how large could the particle be so that it could be said to have a lattice temperature of zero for a large fraction of the time? ($h = 6.6 \times 10^{-34}$ J s; $k_{\mathrm{B}} = 1.4 \times 10^{-23}$ J K^{-1}.)

98

A plate is opaque except for two long thin slits that cross at right angles. The plate is illuminated normally by a parallel beam of monochromatic coherent light. Describe the interference fringes that would appear on a distant screen centred on the line produced by extending the incident beam through the point where the slits cross, and orientated parallel to the plate.

99

The mass of the sun is 2×10^{30} kg (it contains 10^{57} protons), and its radius is 7×10^{8} m. Suppose the sun were to collapse into a neutron star and, in so doing, not change its angular momentum or the total magnetic flux through its equator. Estimate

(i) the decrease in its period of rotation;
(ii) the increase in its surface magnetic field;
(iii) the change in its magnetic field at large distances.

100

A vertical conducting wire of radius 1 mm is coaxial with an earthed cylinder of internal radius 1 cm and height 20 cm. In the air between the two are randomly distributed spherical particles of aluminium dust which, in the absence of electric fields, would take several minutes to settle. Calculate the fraction of the particles reaching the bottom of the cylinder when the wire is raised to a potential of 10 kV. (The density of aluminium is $2.5\ \mathrm{g\,cm^{-3}}$.)

101

On a clear night an observer on the sea shore sees a reflection of the moon in the sea. The image is imperfect because the sea is not perfectly smooth. In addition to this imperfect image, he notes that there is a blaze of light which is brightest at the horizon itself. Give a quantitative explanation as far as you can.

102

A vertical tube, of circular cross section, is sealed at the lower end and filled with a viscous liquid. The tube also contains a cylindrical slug, whose diameter is very slightly less than the bore of the tube, and whose length is two or three times its diameter. The slug falls under the influence of gravity. Derive an equation relating the rate of fall and the viscosity of the liquid.

103

Estimate the maximum number of razor blades which can float, one on top of the other, on the surface of water. (Hint: the maximum depression of the water surface is of the order of the 'surface tension depth' $h = [\text{surface tension}/(\text{density of water} \times g)]^{1/2} = 2.7$ mm.)

104

How would you apply the Lorentz transformation to determine the velocity of propagation of light in a moving medium?

Water, with refractive index 1.33, is made to flow in opposite directions in two parallel pipes, each 20 m long. Parallel light is split into two beams. One beam goes into the first pipe and returns along the second; the other beam enters the second pipe and returns along the first. On re-emerging from the pipes, the beams are recombined and their interference fringes are observed. Assuming the wavelength of the light to be 5×10^{-5} cm, what is the fringe shift when the speed of the water is changed from zero to 10 $m\,s^{-1}$?

105

A parabolic mirror is to be made to focus the sun's disc into a circle of radius 1 cm. Estimate the smallest diameter of such a mirror if it can be used to melt iron. Make any assumptions that are plausible and necessary. Stefan's constant is $\sigma = 5.67 \times 10^{-8}$ $J\,K^{-4}\,m^{-2}\,s^{-1}$; the melting point of iron is 1535°C; the solar constant at the bottom of the atmosphere is 1 $kW\,m^{-2}$.

106

A lake is initially at 0°C. How thick will the ice layer be on the lake if its surface temperature is reduced by 10°C and held at this temperature for 40 days? You may neglect heat exchanges between the lake and the underlying rock but you should discuss quantitatively the validity of neglecting the specific heat of ice, σ, in comparison with the latent heat of freezing, L. The thermal conductivity of ice is 2.3 $J\,m^{-1}\,s^{-1}\,K^{-1}$; $L = 3.3 \times 10^5$ $J\,kg^{-1}$; $\sigma = 2.1 \times 10^3$ $J\,kg^{-1}\,K^{-1}$; the density of ice is 920 $kg\,m^{-3}$.

107

A spherical soap bubble, carrying no electric charge, has a radius r. If electric charge is added to the bubble, it will expand. Explain why. Show that the charge that would be needed to expand the bubble to a radius $2r$ would be

$$8\pi\,[\epsilon_0 r^3 (12\gamma + 7pr)]^{1/2}$$

where γ is the surface tension of the liquid (assumed constant) and

p is the atmospheric pressure. Assume that the expansion is isothermal and that the gas is ideal.

108

A simple model of an insect eye consists of a spherical array of closely packed conical receptors, each optically isolated from its neighbours. Each cone has a small lens which focuses parallel axial light onto a photosensitive cell within the cone. Show that if the eye is simultaneously to achieve maximum sensitivity to light *and* maximum angular resolving power, the sensitive cell should lie at a depth $R/3$ behind the lens, where R is the radius of the spherical array. Estimate the diameter of a single lens in the optimised eye and the minimum resolvable angle between two point sources of green light ($\lambda = 500$ nm) if $R = 1$ mm.

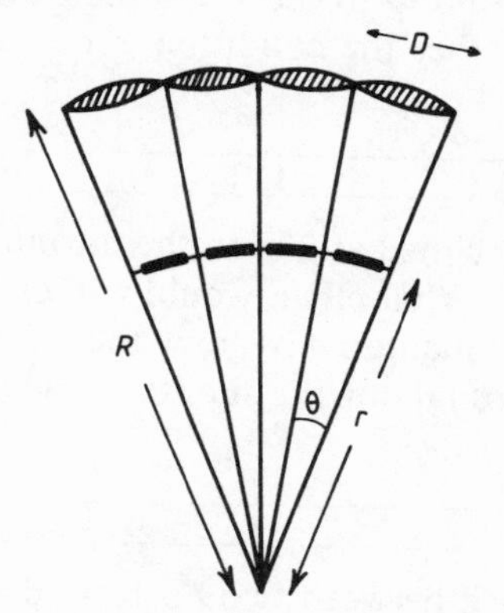

109

An element A (atomic weight 55.85) crystallises with the body-centred cubic structure, with a lattice parameter of 2.861 Å. An alloy is known from chemical analysis to contain 5.00 atomic per cent of B (atomic weight 10.82), the remainder being A. An x-ray powder photograph using radiation of wavelength 1.539 Å gives a reflection which is identified as [110], at a Bragg angle of $22°15'$. The density of the alloy is 7.85×10^3 kg m^{-3}, while that of pure A is 7.93. Determine whether the B atoms are present in substitutional or interstitial solute positions. Avogadro's number is 6.025×10^{23}; $e = 1.60 \times 10^{-19}$ C; $h = 6.625 \times 10^{-34}$ J s. (NB. Some of the information given is redundant.)

110

The drag force on a body of frontal area A moving at velocity v through a medium of density ρ is a function of these three variables multiplied by a dimensionless factor C_D. If C_D may be taken to be constant over the range of velocities involved, and the atmospheric air density varies with height h according to $\rho = \rho_0 \exp(-h/h_0)$, at what height is the maximum aerodynamic drag on the space shuttle encountered if the vehicle is launched vertically and achieves a constant acceleration?

111

A hoop of radius r is thrown, with its plane vertical, with horizontal velocity v_0 and retrospin ω_0. Show that ω_0 must be greater than v_0/r in order for the hoop to return to the thrower on hitting the ground. Discuss whether the condition is the same for a thin circular disc.

112

The earth's axis is inclined at 23° to the normal through the plane of the moon's orbit. What effect would you expect this to have on the tides? Explain qualitatively how the effect varies (a) with latitude and (b) throughout the month. Neglect the effect of the sun.

113

A soap film is formed between two horizontal coplanar concentric wire rings whose radii are $3a$ and a. The outer ring is then raised, and the inner ring hangs in equilibrium a distance d below it, supported by the film. Show that d cannot exceed $1.76a$.

Note:

$$\frac{\mathrm{d}}{\mathrm{d}x}\{\ln[cx + (c^2x^2 - 1)^{1/2}]\} = c(c^2x^2 - 1)^{-1/2}.$$

114

A drunkard performs a random walk but, as a result of increasing fatigue, each step taken is shorter than the previous one by a factor f, where $f < 1$. If the initial step was of length λ, what is the RMS distance from the starting point after N steps?

115

Suppose a system of two parallel slits is set up, illuminated by a plane monochromatic light wave in order to produce widely separated interference fringes on a distant screen. If an eye is placed (*a*) at one of the dark fringes, (*b*) at one of the bright fringes, what will be seen on looking towards the slits?

116

A neutron interferometer splits a monoenergetic beam of neutrons into two parallel beams, a distance d apart. These are brought together after each has travelled a path length l. It is found that if the two beams are in the same horizontal plane the interference is constructive, but if the system is rotated so that one beam is vertically above the other the interference becomes destructive. Explain this result and obtain an expression for the neutron energy in terms of d, l and any relevant general constants.

117

A thin jet of water (mean diameter 0.1 cm) flows vertically from a tap to the sink 20 cm below at a flow rate of 2 $cm^3\,s^{-1}$. An electrified rod (a ball-point pen, rubbed on a jacket, does very well) is held horizontally, so that its mid-point is 5 cm below and 1 cm to one side of the tap. It is found that the jet of water is deflected, and now falls in the sink 3 cm away from the point below the tap.

Estimate the charge per unit length on the electrified rod, treating the water as a non-conducting liquid of relative permittivity 81.

118

Two parallel radiating surfaces are at temperatures T_1 and T_2. The emissivities are ε_1 and ε_2 respectively and $T_1 < T_2$. Show that the net radiative flux of energy from the hotter to the colder surface is

$$\frac{\sigma(T_2^4 - T_1^4)\varepsilon_1\varepsilon_2}{\varepsilon_1 + \varepsilon_2 - \varepsilon_1\varepsilon_2}$$

where σ is Stefan's constant.

A thin metal screen of emissivity ε is now placed between the original two surfaces to act as a radiation shield, and is allowed to take up some temperature T. By what factor will this radiation shield reduce the net radiation flux between T_2 and T_1 if $\varepsilon = \varepsilon_1 = \varepsilon_2$?

Will the use of more than one shield reduce the radiation flux still further?

119

---

*A crude estimate of the depth of a well can be obtained by dropping a stone down it and listening for the splash, then using $s=\frac{1}{2}gt^2$. You are asked to make first-order estimates of the corrections due to

(i) the finite velocity of sound ($v_s = 340\ \mathrm{ms^{-1}}$),
(ii) air drag.

Which of these is the bigger correction for $s = 30$ m and a stone of radius $a = 2$ cm and density $3\times10^3\ \mathrm{kg\,m^{-3}}$? Would your conclusion be significantly different if a table tennis ball were dropped in the same way ($a = 2$ cm, $m = 1.7$ g)? You may assume that the splash would still be audible! (The drag for a sphere of radius a depends on the dimensionless Reynolds number $R = \rho va/\eta$, where ρ is the fluid density and η is its viscosity. For $R < 10$, Stokes' law $F = 6\pi a\eta v$ is valid, but for $R > 100$ the drag F is given by $F = C_D\rho\pi a^2v^2$; the variation of C_D with $\log_{10}R$ is given in the diagram. For air at 20°C, $\rho = 1.2\ \mathrm{kg\,m^{-3}}$ and $\eta = 1.8\times10^{-5}\ \mathrm{kg\,m^{-1}\,s^{-1}}$.)

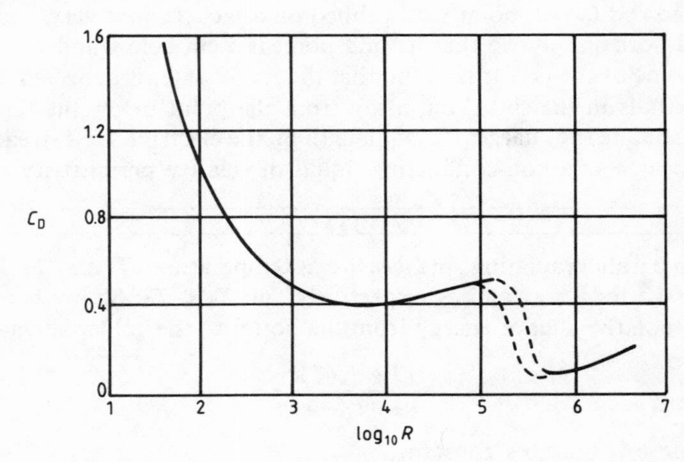

120

A cosmic-ray electron is moving in essentially free space near the earth. At $t = 0$ it lies in the equatorial plane of the earth's magnetic field, and its velocity vector v also lies in this plane. Discuss qualitatively the resulting motion of the electron. If its speed is such that the radius of curvature of its path is small compared with its distance from the centre of the earth, show that it will make a complete circuit of the earth in a time of the order of $R^2\omega/v^2$, where R is its mean distance from the centre of the earth and ω is the cyclotron frequency ($\omega = eH/m$).

121

The latent heat of vaporisation of water is $H = 2.25 \times 10^9$ J m^{-3}. The surface tension of water is $T = 72$ mN m^{-1}. From these two facts, together with a simple theory, estimate the mean distance between neighbouring water molecules. (Hint: introduce the bond energy E between two water molecules.)

122

Kinetic theory suggests that, at least for dilute gases, the viscosity η is independent of the pressure. Suppose that the force between molecules is λ/r^n, then use dimensional analysis to obtain the relationship between the viscosity η, the molecular mass m and the mean speed V.

Data for oxygen suggest that the viscosity varies with temperature T as $\eta = T^{0.725}$. Estimate the value of n.

123

A circular hoop made of flexible elastic material is made to spin in its own plane with circumferential speed v.

(i) If a small disturbance to the shape is made instantaneously at one point of the hoop, what will happen?

(ii) As a result of the centrifugal force the hoop (now circular again) becomes enlarged by a fraction f. Show that $f = (v/c)^2$, where c is the velocity of longitudinal elastic waves.

(iii) A length of wire lying on the surface of the earth in an east–west direction could be regarded as part of a hoop around the earth. Does the above result imply that the rotation of the earth stretches it by a significant amount?

124

A metal sphere revolves around the earth in a circular orbit, so that it is always in sunlight. If the emissivity of its surface may be chosen to vary with wavelength in any desired manner, estimate the lowest equilibrium temperature possible for the sphere when its orbit is outside the atmosphere.

The sun subtends an angle of 0.5° at the earth, and radiates as a black body at 5700 K. The earth may also be regarded as a black body.

125

In a nuclear chain reaction the number of neutrons present as a function of time, $N(t)$, follows an equation of the form

$$T\frac{\mathrm{d}N}{\mathrm{d}t} = -CN(t) + N(t\text{-}\tau).$$

T is a small time parameter ($\sim 10^{-7}$ s) and τ (the delayed neutron delay time) is large (~ 10 s). C is a constant of order one.

(*a*) Show that solutions are of the form $N(t) = N_0 \exp(\alpha t)$ where N_0 and α are constants, and find an equation for α.

(*b*) Show that (i) if C is negative then the exponential growth has a time scale of the order of microseconds and (ii) if C is positive but less than 1, the growth has a time scale of seconds.

(*c*) What happens if C is greater than one?

126

The diagram represents a section of a capacitor whose rectangular plates are inclined at an angle ϕ. The plates are of length z, perpendicular to this section. Show that, ignoring edge effects, the capacitance is

$$\frac{\epsilon\epsilon_0 z}{\phi}\ln\left(1+\frac{a}{b}\right)$$

where ϵ is the relative dielectric constant of the medium between the plates.

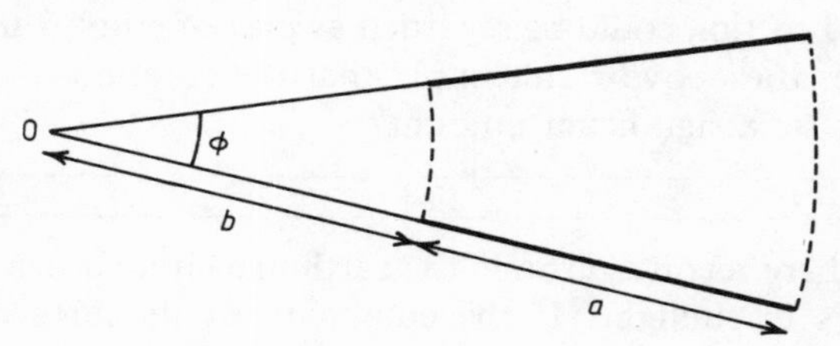

127

When a wire is stretched from length l to $l + \mathrm{d}l$ its electrical resistance is observed to increase from R to $R + \mathrm{d}R$. It is said that Poisson's ratio is given by

$$\nu = \frac{1}{2}\left(\frac{l}{R}\frac{\mathrm{d}R}{\mathrm{d}l} - 1\right).$$

What dubious assumption underlies this calculation?

128

A man jumps as high as possible on a small planet. Estimate how small it would have to be for him to be able to jump off altogether. (The radius of the earth is 6380 km.)

129

It has been stated that Archimedes once saved the Greeks from attack by a Roman fleet by equipping a large number of well trained soldiers with large hand-held plane mirrors and getting them to reflect the sun's rays onto the same part of a Roman ship at a distance of 100 m, thereby setting it on fire. Comment on the feasibility of such a plan. (You may assume that the mirrors are available, that the solar flux is $1\ \mathrm{kW\,m^{-2}}$, and that the angular diameter of the sun is $0.5°$. The ships may be assumed to be wooden. Stefan's constant $\sigma = 5.7 \times 10^{-8}\ \mathrm{W\,m^{-2}\,K^{-4}}$.)

130

An experimenter performs a version of the Cavendish experiment and attempts to measure the force of attraction between two massive earthed conducting spheres. She neglects to shield her apparatus from stray fields. What effect, if any, would uniform electric and magnetic fields perpendicular to the line joining the spheres have on her measurement of the gravitational constant G?

131

My bottle of soy sauce has a cap with a small hole in it—about 2 mm diameter. When I turn it upside down, a small quantity of liquid runs out and the flow stops. If I now turn it the right way up and then invert it again, some more runs out; and so on. Explain this behaviour. Your discussion should be quantitative and should lead, after suitable approximations, to a simple algebraic expression for the amount discharged, in terms of the other parameters involved. In order to avoid unnecessary complications, make whatever simplifying assumptions you think fit about the shape of the bottle.

132

A thin copper ring rotates freely about a diameter, which is perpendicular to a uniform magnetic field B. Its initial frequency of rotation is ω. Calculate the time taken for the frequency to decrease to $1/e$ of its original value, under the assumption that the energy goes into Joule heat. (The resistivity of copper is $1.8 \times 10^{-8}\ \Omega\,\mathrm{m}$; the density of copper is $8.9 \times 10^3\ \mathrm{kg\,m^{-3}}$. Take $B = 2 \times 10^{-2}$ T.)

133

A cylindrical tube, about one metre long and about a tenth of that in diameter, is open at both ends. It is in, and initially filled with, air at atmospheric temperature and pressure in which the velocity of sound is C_0, say. A thin piston is driven through the tube from one end at a speed $2C_0$. Discuss the density of the air just behind the piston when it is half way through the tube, and obtain an estimate of its value.

134

The density of solid argon (atomic weight 40) is $1.6 \times 10^3\ \mathrm{kg\,m^{-3}}$. Estimate the thermal conductivity of argon at STP.

135

An experiment to measure the viscosity of a liquid uses a rotation viscometer with cylinders of radii a and b, where $a, b \gg (a-b)$, one of which rotates; the axes of the cylinders are parallel but not quite coincident, the distance between them being $x \ll (a-b)$. Show that the first-order correction term in x is zero and find an expression for the second-order term.

136

Since parallel current filaments attract one another, one might think that a current flowing in a solid rod of circular cross section would tend to concentrate near the axis of the rod. Does this happen? If so, estimate the size of the effect. Would it be detectable?

(The conduction electron density in a typical metal is of the order of 10^{29} m^{-3}. The electronic charge is 1.6×10^{-19} C.)

137

It takes about four minutes to boil a 2 oz hen's egg, to be acceptable to most people for eating. For how long would it be advisable to boil an ostrich egg, which weighs about 48 oz?

Part 2

The Solutions

1

The total volume of the depression in the water surface must, by Archimedes' principle, be 8.5 times that of the steel wire. So, if the wire is to float as described, the depression must be much wider (approximately 8.5 times wider) than it is deep. The slope of the water surface may therefore be considered small. If $y(x)$ denotes the depression at horizontal position x, with the wire centre at $x=0$, then the curvature of the surface may be taken as $\mathrm{d}^2y/\mathrm{d}x^2$ (since $\mathrm{d}y/\mathrm{d}x \ll 1$). We thus have $\gamma\, \mathrm{d}^2y/\mathrm{d}x^2 = \rho g y$ where γ is the surface tension and ρ is the density of water. The solution of this which decays to zero at $x=\infty$ is

$$y = y_0 \exp[-(\rho g/\gamma)^{1/2}|x|].$$

The area of the depression is then

$$2\int_0^\infty y\, \mathrm{d}x = 2y_0(\gamma/\rho g)^{1/2}.$$

This expression ignores the (small) correction arising from the fact that near $x=0$ the shape of the water surface will be given by the shape of the wire, rather than by the above equation. Archimedes' principle then gives $\pi r^2 8.5 \rho g = \rho g 2 y_0 (\gamma/\rho g)^{1/2}$. Using the numerical values above, this gives $y_0 = 0.2$ mm exactly. Because of the same correction previously mentioned, the centre of the wire will be slightly above the level of the water surface.

2

Consider a sphere of density ρ, radius r and mass M $(=\frac{4}{3}\pi r^3 \rho)$. The gravitational energy released by adding a further shell, density ρ and thickness δr, using matter originally at infinity is $(GM/r)4\pi r^2 \rho \delta r$. The total energy corresponding to a homogeneous sphere of radius r_0 is given by integrating this quantity from zero to r_0. The result is $3GM^2/5r_0$. This result applies both to the original dust cloud, radius R, if this is assumed to be homogeneous, and to the final planet, radius a. The difference between the two expressions thus gives the energy released when the one collapses into the other, whatever the details of the intermediate stages. Thus

$$E = \frac{3GM^2}{5}\left(\frac{1}{a} - \frac{1}{R}\right) \simeq \frac{3GM^2}{5a} \qquad \text{since } R \gg a.$$

This energy is available for heating the matter, and so can be equated to $Ms\Delta\theta$ (s = specific heat, $\Delta\theta$ = temperature rise). Hence

$$a^2 = \frac{5}{4\pi}\frac{s\Delta\theta}{G\rho}.$$

If we now estimate the values $s = 10^3\ \mathrm{J\,kg^{-1}\,K^{-1}}$, $\Delta\theta = 2000\ \mathrm{K}$, $\rho = 6\times10^3\ \mathrm{kg\,m^{-3}}$, we obtain $a = 1.40\times10^6$ m.

The assumption that ρ = constant is certainly not valid, but a more realistic approach would be complicated, and unlikely to give very different results. The neglect of radiation losses is also unrealistic and would lead to temperatures that were too high. On the other hand, in the real world, radioactive heating would be important and would work in the opposite direction. We may note that the result $a \simeq 10^6$ m is smaller than the radius of the planets in the solar system, which have therefore probably been molten at some stage, but larger than many asteroids, which have therefore probably not.

3

The dominant consideration is the conversion of the kinetic energy of the running man to gravitational potential energy with something less than 100% efficiency. If, in his approach, he attains a speed of 10 m s^{-1}, the corresponding rise is about 5 m. Smaller terms arise from (i) the fact that his centre of mass is already about 1 m above the ground when he starts; (ii) the work done by his legs on take-off, and by his arms in climbing up the pole (? $\simeq$ 0.5 m); (iii) the fact that his centre of mass actually passes *below* the bar (? $\simeq$ 10 cm). Adding these items together gives an answer of approximately 21 feet. The difference between this and the observed 18 feet is due to an efficiency of less than 100%—or to errors in the estimated quantities. In any case, there is clearly no hope of Mr Fibreglass making good his boast.

4

Since the viscosity is zero it is tempting to try a solution using the conservation of energy principle. In fact, this approach leads to incorrect answers (see below) unless we take into account internal motions in the liquid, just as in the problem of the punctured soap film (problem 48). Instead, we must use Newton's law in the form applicable to a system of variable mass. If we denote the radius by

a, the cross sectional area of the tube by A and the height of the column at time t by x, we have

$$\frac{\mathrm{d}}{\mathrm{d}t}(\rho A x v) = 2\pi a\gamma - \rho A x g = \rho A X g - \rho A x g$$

where $X \equiv 2\pi a\gamma/A\rho g$ is the equilibrium height of rise of the liquid. This gives $\mathrm{d}(xv)/\mathrm{d}t = g(X - x)$. By making the substitution $xv = z$, this can be written as $z\mathrm{d}z/\mathrm{d}x = g(Xx - x^2)$, which can be integrated to give

$$v^2 = g(X - 2x/3 + b/x^2) \tag{1}$$

where b is an arbitrary constant. It is not possible to choose b to satisfy the reasonable initial condition $v = 0$ at $x = 0$. A refinement of the model, which also circumvents this difficulty, is to take some account of the motion of the liquid in the reservoir, by adding a small (constant) amount to the mass of liquid moving in the tube. This can be done by replacing x by $(x + x_0)$ in the previous argument, to give

$$v^2/g = (X + x_0) - 2(x + x_0)/3 + b/(x + x_0)^2.$$

If we now put in the initial condition $v = 0$ at $x = 0$, we obtain

$$v^2/g = (X + x_0/3 - 2x/3) - x_0^2(X + x_0/3)/(x + x_0)^2$$

which can be written in dimensionless form as

$$\frac{v^2}{gX} = \left(1 + \frac{1}{3}\frac{x_0}{X}\right)\left[1 - \left(\frac{x_0/X}{x/X + x_0/X}\right)^2\right] - \frac{2}{3}\frac{x}{X}.$$

This curve has been plotted for $x_0/X = 0.1$ and 0.01 in diagram (*a*) below. It will be seen that $v^2 = 0$ at $x = 0$ and near $x/X = 3/2$. (There is a third root near $-2x_0/X$ which is of no physical interest.) Near both extremes of the motion, v^2 depends almost linearly on x, which implies that $x = kt^2$, with the value of k proportional to the slope of the line. We can thus sketch the graph of x against t, which consists of parts of two parabolas, joined smoothly (diagram (*b*)). The motion is thus oscillatory, but not simple harmonic.

If we had applied the conservation of energy principle, we would have obtained

$$\frac{v^2}{gX} = \frac{2(1 - x/2X)(x/X)}{x/X + x_0/X}.$$

This gives $v^2 = 0$ at $x = 0$ and $x = 2X$; the resulting curves are qualitatively similar but differ in details.

Within the limitations of the model used, the motion will be undamped, since there is no mechanism for energy dissipation. Denote by V_c the critical velocity above which the viscosity is non-zero and draw the line AB on diagram (*a*) at a height corresponding to this. From the origin up to point 1 the motion will be unaffected. From 1 to 2 the velocity will fall progressively below the value for the undamped motion, as shown schematically by the broken curves in diagram (*a*). From 2 to 3 the motion will be undamped, and therefore reversible, but from 2 to 4 more energy will be dissipated. Clearly the excursions of the meniscus will become pro-

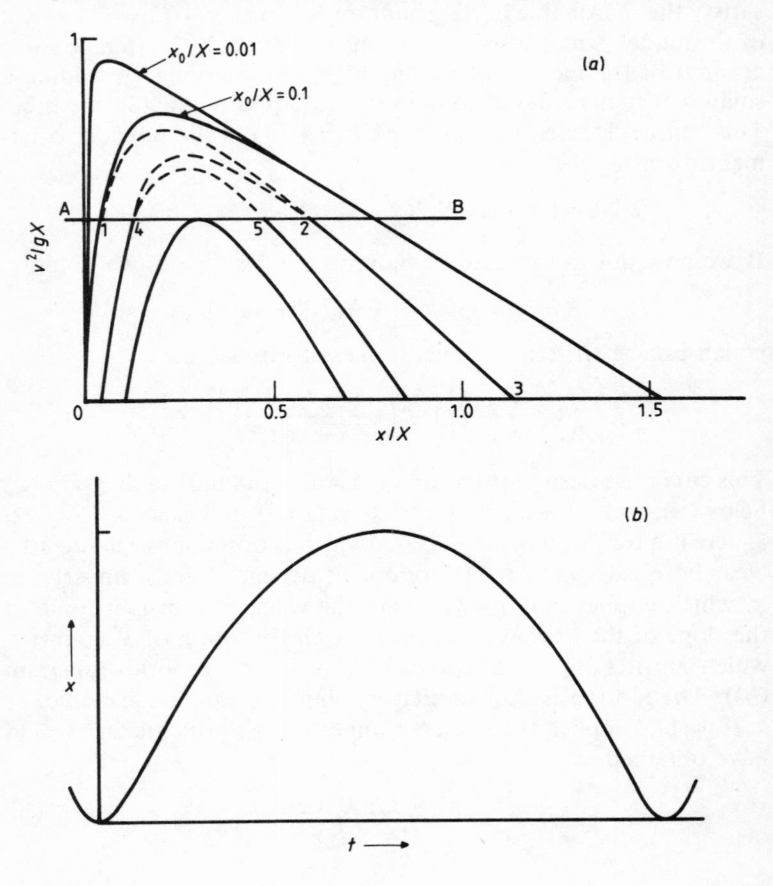

gressively smaller at both ends, until a point is reached when the maximum velocity is less than V_c. Thereafter the motion will be undamped.

5

We have $(p_a - P) = 4S/a$ and $V_a = 4\pi a^3/3$, with similar expressions involving b and c. Since the total mass of air inside is conserved, we also have $p_a V_a + p_b V_b = p_c V_c$. This gives

$$S = -\frac{P}{4}\left(\frac{a^3 + b^3 - c^3}{a^2 + b^2 - c^2}\right).$$

For a discussion of the errors involved in this method of finding S, we may simplify the algebra by writing $b = a$, without altering the conclusions. Hence

$$S = \frac{P}{4}\frac{2a^3 - c^3}{2a^2 - c^2}.$$

But c^3 will be approximately equal to $2a^3$—since the initial and final pressures will both be nearly equal to P. Hence in the discussion of errors we can write $c^3 = 2a^3$ in the denominator, and confine our attention to the errors arising from the numerator. We thus write $4S(2a^2 - c^2)/P = Q(\text{say}) = 2a^3 - c^3$, and so $\delta Q = 6a^2\delta a - 3c^2\delta c$. We then set $|\delta a| = |\delta c|$, but since these are independent observations we must write $\delta Q = (36a^4 + 9c^4)^{1/2}\,\delta r$, where δr is written for δa or δc. This gives

$$\delta S = \frac{P}{4}\frac{7.66a^2\delta r}{2a^2 - c^2}.$$

Thus if we write $c^3 = 2a^3$, $\delta S = 4.64P\,\delta r$. Very careful measurements might reduce δr to 10^{-4} cm. In CGS units $P \simeq 10^6$, and so $\delta S \simeq 500$ dyn cm^{-1}, while S itself is quoted as approximately 30 dyn cm^{-1}. Clearly the method is not very accurate.

6

Since those decay products which are pions will themselves decay with short lifetimes, and the positrons will be rapidly annihilated, we can assume that almost the whole mass of the nucleons will, eventually, be released as energy. (A small proportion of the energy will be lost in decays which involve neutrino production, but detailed calculations suggest that the energy lost in this way will only be approximately 10% of the total—which is a negligible correction in a calculation of this kind.) Thus m kg of body mass will produce an energy $E_0 = mc^2 = 9 \times 10^{16}\, m$ J. Since this total energy will decay as $E_0 \exp(-t/\tau)$, the initial rate of energy production is E_0/τ J/year, if τ is in years. Thus a 20-year-old student weighing m kg will have received about $20 \times 10^{17}\, m/\tau$ J—assuming that only a small fraction of the energy produced escapes from her body, which is reasonable. This amounts to $2 \times 10^{18}/\tau$ J kg^{-1} $= 2 \times 10^{20}/\tau$ rads. The fact that she is still alive implies that $2 \times 10^{20}/\tau < 600$, i.e. $\tau > 3 \times 10^{17}$ years. A more careful estimate would allow for the fact that not all of the m kilograms have existed for the whole 20 years.

To raise this lower limit it would be necessary to construct apparatus which would detect individual decay events—e.g. using photomultipliers. A 'source' consisting of M kg of water would contain about $6 \times 10^{26}\, M$ nucleons, which would decay giving rise to $6 \times 10^{26}\, M/\tau$ events per year. If the equipment could detect one event per year with confidence, then the limit for τ is $6 \times 10^{26}\, M$ years. The upper limit on M would depend on the outcome of a cost-benefit analysis.

7

The equation of motion is $m\ddot{x} = -cx^3$. If we multiply both sides by $2\dot{x}$ and integrate, we obtain $m\dot{x}^2 = c(x_0^4 - x^4)/2$, if $\dot{x} = 0$ at $x = x_0$. Writing this as $(m\dot{x})^2 + mcx^4/2 = mcx_0^4/2$, we see that the graph of momentum $(m\dot{x})$ against position (x) is a closed loop. The motion is thus periodic. For SHM the loop would be a circle. For this motion, it is much flattened where it crosses the p axis, showing that p remains almost constant for a considerable range of x on either side of the axis—i.e. the curve of x against time is more nearly linear where it crosses the time axis than is a sine curve.

The above equation can be written $(\mathrm{d}t/\mathrm{d}x)^2 = 2m/c(x_0^4 - x^4)$. If we set $t = 0$ at $x = 0$, then at $x = x_0$, $t = \tau/4$ (τ is the period). The equation can then be integrated to give

$$\int_0^{\tau/4} \mathrm{d}t = \left(\frac{2m}{c}\right)^{1/2} \int_0^{x_0} \frac{\mathrm{d}x}{(x_0^4 - x^4)^{1/2}}.$$

If we write $x/x_0 = u$, this becomes

$$\frac{\tau}{4} = \left(\frac{2m}{c}\right)^{1/2} \frac{1}{x_0} \int_0^1 \frac{\mathrm{d}u}{(1-u^4)^{1/2}} = \left(\frac{2m}{c}\right)^{1/2} \frac{k}{x_0}$$

where k is a numerical constant which is clearly somewhat greater than unity. Its value can be computed: it is equal to 1.31. Thus $\tau x_0 = 5.2(2m/c)^{1/2}$, which is the required result. The numerical value is 2.3 s for $x_0 = 1$ cm.

8

We can treat this solenoid as being infinitely long. Let r_1, r_2 denote the inner and outer radii of the windings, l the length, ρ the resistivity, j the current density and B the field.

(i) Approximate the spiral by a series of close-fitting cylinders, each of thickness dr. The field due to any one of them is $\mathrm{d}B = \mu_0 j \mathrm{d}r$ and the total field $B = \mu_0 j(r_2 - r_1)$. The power dissipated is $\int j^2 \rho \, \mathrm{d}V$, where d$V$ is an element of volume. In this case this is simply $j^2 \rho V$, where $V = \pi(r_2^2 - r_1^2)l$, and j is given by the above expression for B. This gives the power

$$W = \frac{\pi B^2 \rho l}{\mu_0^2} \frac{r_2 + r_1}{r_2 - r_1} = 597 \text{ kW}.$$

(ii) The current density will now depend on the radius in such a way as to make $j 2\pi r$ a constant, A, related to the EMF around one disc. Thus we now have $\mathrm{d}B \equiv \mu_0 j \, \mathrm{d}r = \mu_0 A \, \mathrm{d}r/2\pi r$, and so $B = \mu_0 A \ln(r_2/r_1)/2\pi$. The power dissipated is again given by

$$\int j^2 \rho \, \mathrm{d}V = \int_{r_1}^{r_2} \left(\frac{A}{2\pi r}\right)^2 \rho l \, 2\pi r \, \mathrm{d}r$$

where A is given by the above expression for B. In this case the power is

$$W = \frac{2\pi B^2 \rho l}{\mu_0^2} \frac{1}{\ln(r_2/r_1)} = 494 \text{ kW}.$$

(For a long solenoid consisting of a single layer of wire, the internal field is independent of the radius. It follows that, in this problem, if the total current flowing around the solenoid is fixed, the resulting field is independent of the radial distribution of current density. The design problem is thus to choose the distribution that minimises the energy dissipation. Intuitively it would seem that this would be achieved by arranging for a larger current density near the inner radius, where the resistive path length is shorter. Formally, we must minimise $\int_{r_1}^{r_2} j^2(r) r \, \mathrm{d}r$ subject to $\int_{r_1}^{r_2} j(r) \, \mathrm{d}r = \text{constant}$.

The solution is $j \propto 1/r$, but the mathematics is difficult. This is, in fact, the distribution in case (ii) above.)

9

If the cross section is $2a \times 2b$ ($a \ll b$), the viscous flow can be treated as if between two infinite flat plates. The standard treatment gives, for the velocity profile, $v = P(a^2 - x^2)/2\eta l$, where x is the distance from the centre plane and P/l is the pressure gradient. The volume rate of flow is thus $\dot{V} = 4ba^3P/3\eta l$. In the capillary tube, when the liquid has reached a height h, we can write $l = h$, $P = (T/a - \rho g h)$(T is the surface tension, ρ is the density) and $\dot{V} = 4ab\ \mathrm{d}h/\mathrm{d}t$. If we make these substitutions, we obtain

$$\frac{\mathrm{d}h}{\mathrm{d}t} = \frac{a^2}{3\eta}\frac{T/a - \rho g h}{h} = \frac{\rho g a^2}{3\eta}\frac{h_0 - h}{h}$$

where $h_0 \equiv T/\rho g a$ is written for the equilibrium height of rise. After integrating, with $h = 0$ at $t = 0$, we have

$$\frac{\rho g a^2}{3\eta}t = h_0 \ln\left(\frac{h_0}{h_0 - h}\right) - h.$$

Putting $h = 0.99\, h_0$, we obtain $t = 6.5 \times 10^7$ s ($\simeq 2$ years!).

Notes. (i) The argument ignores the force necessary to accelerate the fluid from rest, i.e. 'inertia forces' are neglected compared with 'viscous forces'. This is justified *a posteriori* by calculating the Reynolds number $R \equiv \rho v l/\eta$, which is found to be much less than unity. (ii) The velocity profile given is clearly wrong near the advancing surface, which must remain concave upwards, but this will affect to any appreciable extent only a length of the column short compared with h_0 ($\simeq 1.5 \times 10^3$ cm!). (iii) The equation gives $\mathrm{d}h/\mathrm{d}t = \infty$ when $h = 0$, but this will affect only the very early stages and will have little effect on the final result.

10

(i) The governing equation is $M\ddot{z} = c - k\dot{z}$ where $M = 2\pi\sigma a^3/3$, $c = 4\pi a^3 \sigma g/3$ and $k = 4\pi\eta a$. The solution, with $z = \dot{z} = 0$ at $t = 0$, is

$$z = 2g\tau\{t - \tau[1 - \exp(-t/\tau)]\}$$

where $\tau = \sigma a^2/6\eta$.

If $t/\tau \ll 1$, an approximation for z is $z = gt^2$. For the example given, $\tau = 0.042$ s. To obtain a numerical solution, the equation can be rewritten as

$$t = (\tau + z/2g\tau) - \tau \exp(-t/\tau)$$

which lends itself to solution by iteration. If $z = 5$ cm, $t = 0.099$ s; if $z = 10$ cm, $t = 0.162$ s.

For part (ii) write the equation as $\ddot{z} + (k/M)\dot{z} - c/M = 0$. Then c/M does not involve a, and $k/M \propto 1/a^2$. Thus as the bubble rises and gets larger, the effective drag force will fall and the bubble will move faster than it would if a remained constant. The bubble will also cease to be even approximately spherical: it becomes mushroom shaped.

11

At a point P in the tunnel, a distance x from its mid-point, let the acceleration due to gravity be g. Then the component along the tunnel is $f = g\cos\theta = gx/r$. If we assume that the density of the earth is uniform (which is not true!) it is easy to show that $g = g_0 r/r_0$, where g_0 is the value of g on the surface, i.e. at $r = r_0$. Thus $f = g_0 x/r_0$ and the motion is simple harmonic, with period $\tau = 2\pi(r_0/g_0)^{1/2}$. The time from A to B is thus $\tau/2 \simeq 42$ minutes.

Apart from the prohibitive initial construction costs, problems due to friction, air resistance, the increased temperatures at depth and, eventually, the liquid core, this would be quite an attractive scheme!

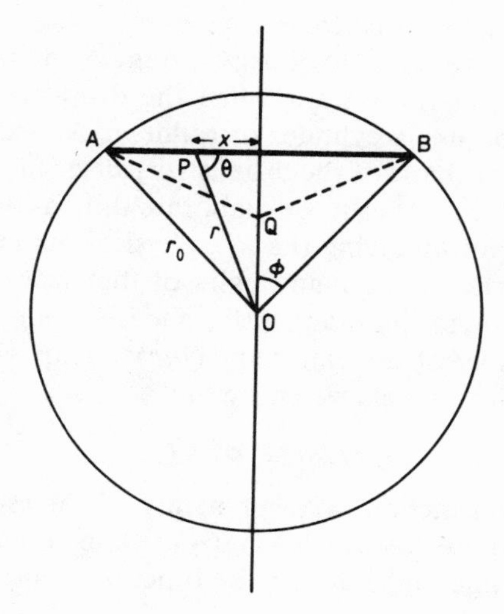

No, the straight tunnel does not provide the quickest path. It is advantageous to travel along a symmetrical curved path which approaches closer to the centre than the straight line. The relevant calculations are difficult, but it is comparatively simple to show that it is even advantageous to go along the path AQB in the diagram, and that the optimum position of Q is when AQ bisects the angle OAB. The time for the journey is then $2t$, where $\cos 2\pi t/\tau = (1 - \sin \phi)/(1 + \sin \phi)$ and the angle AOB $= 2\phi$.

12

The physical constants likely to have an influence on this phenomenon are ρ, the density of the liquid, γ, its surface tension, and g, the acceleration due to gravity. The observation must be made slowly, i.e. we suppose that we are dealing with a series of equilibrium situations, so that the viscosity is not involved. The standard method of dimensional analysis shows that the only combination of ρ, g and γ which has the dimensions of a length is $a = (\gamma/\rho g)^{1/2}$, a^2 being the so-called capillary constant.

The most elementary physical analysis considers a drop whose surface, where it joins the solid tube, has the form of a short cylinder of radius r, where r is the radius of the tube. If p denotes the pressure in the liquid at the level of the end of the tube, then the equilibrium of the drop below this level requires $mg + p\pi r^2 = 2\pi r\gamma$. Since, at this level, the liquid surface is supposed to be cylindrical, $p = \gamma/r$ and hence $mg = \pi r\gamma$.

However, this model is obviously wrong. A moment's observation shows that, at the instant when the drop breaks loose, the surface is nothing like a cylinder of radius r. An exact calculation of the equilibrium shape of the drop is very difficult, but fortunately not necessary. A rather more elaborate dimensional analysis of the same kind, but involving the five physical quantities g, ρ, γ, m and r, shows that, of the many pairs of dimensionless quantities which can be formed simultaneously, one is $(\gamma r/mg)$ and $(\gamma/\rho g r^2)$, the second member of the pair being $(a/r)^2$. Thus, without involving any particular model, we can write

$$\gamma r/mg = \phi(a/r)$$

where ϕ is some function. Measurements of corresponding values of m and r on liquids for which ρ and γ are known have shown that, over a considerable range of a/r, the function is almost a constant, giving $\gamma = 0.263\ mg/r$ (instead of $(1/\pi)mg/r = 0.318\ mg/r$ as in the simple theory) and that the 'constant' decreases slowly at both higher and lower values of a/r.

If the drops form quickly, we no longer have a series of quasi-equilibrium situations, and a full solution would involve fluid dynamics. It is clear that we must add to the list of relevant variables the time taken to form the drop and the coefficient of viscosity. The dimensional analysis can still be performed, but the results are not very helpful. It is clear that if the equilibrium theory is to be applied, the time scale of the observations must be increased for liquids of high viscosity.

13

Sound will propagate normally in a gas at reduced pressure, so long as the mean free path of the molecules, l, is small compared with the wavelength λ. At 1 cm Hg, $l \simeq 0.004$ mm, while λ will have its normal value, of the order of 1 m. Thus sound can propagate perfectly well.

The observed effect is due to the acoustic mismatch between the air and the glass. When a plane wave is incident normally on the plane interface between two media A and B, a fraction R of the energy is reflected, given by $R = (Z_A - Z_B)^2/(Z_A + Z_B)^2$, Z being the acoustic impedance: $Z = (\rho E)^{1/2}$, where ρ is the density and E the appropriate elastic modulus. The fraction transmitted is $1 - R$; when Z_A and Z_B are very different this is approximately equal to $4Z_1/Z_2$, where Z_1 is the smaller of Z_A and Z_B, and Z_2 is the larger. For glass, Z is of the order of 10^7 in SI units. For a gas the appropriate elastic modulus is the adiabatic bulk modulus γp, where p is the pressure. For air, Z is about 400 in SI units, and so the fraction transmitted from air to glass (or vice versa) is of the order of 10^{-4}. Since, for a gas, $Z = (\rho\gamma p)^{1/2}$, we see that Z is proportional to the pressure. Thus the fraction of sound energy transmitted will fall as the pressure is reduced.

While this discussion gives some insight into the physical mechanism involved, it is not directly applicable to the problem in hand. We must consider the passage of the wave from the glass into the air again, and also note that the whole process is greatly complicated by internal reflections in the glass. This problem is soluble for plane surfaces and plane waves, but, in fact, neither is plane.

Nevertheless, the general result that the impedance mismatch gets worse as the pressure is reduced remains valid.

An alternative approach is to model the system as two masses (the bell itself and the surrounding glass) connected by a weak spring (the contained air). The amplitude of movement of the bell will, to a good approximation, be constant. The stiffness of the spring, as explained above, will be proportional to the gas pressure. It is then easy to show that the amplitude of vibration of the jar will also be proportional to p, and so the intensity of the sound radiated will fall as p^2.

14

The maximum heat available per unit mass of oil is the chemical 'heat of combustion'.

(i) In a domestic furnace, combustion may not be quite complete (some soot may be formed), but the main heat loss is due to hot gas escaping up the flue, as a result of inefficient heat exchange with the water in the circulating system. The overall efficiency is usually about 70%. Some improvement might be obtained by using the flue gases to pre-heat the incoming air for combustion, but this would increase capital costs.

(ii) In a power station, with a more elaborate installation, the fraction of heat extracted from the flue gases may be somewhat higher. But there is a theoretical upper limit to the efficiency with which this can be converted into mechanical work in the turbines. The thermodynamic efficiency is $(T_1 - T_2)/T_1$, where T_1 is the steam temperature (say 550°C) and T_2 is the cold reservoir temperature (say 70°C). This gives an upper limit to the efficiency of about 60%. In practice, efficiencies of about 40% are more usual, the difference arising mainly from losses in the furnace, as in (i). Electrical losses in the generator, in overhead cables and in transformers account for about a 10% loss. Once the power is through the consumer's meter, the efficiency of conversion to heat is 100%. The overall efficiency of the system is thus not greater than about 36%. Some improvement could be obtained by using the waste heat from the power station, e.g. for district heating schemes.

(iii) The efficiency as far as the consumer's meter is the same as in (ii), but the use of a heat pump can give efficiencies greater than unity, if efficiency is defined as (heat out)/(work in). The Carnot limit is $T_3/(T_3 - T_4)$, where T_3 is the output (room) temperature (say about 20°C) and T_4 is the temperature of the colder (outside) reservoir (say about 0°C). This gives a theoretical limit of about 14 but, in practice, domestic systems usually operate below this by a factor of 2 or 3. Using an efficiency of 5 for the heat pump as a reasonable guess, the overall efficiency of the system is about 1.8.

The order of merit is thus (*a*) heat pump, (*b*) domestic central heating and (*c*) electric radiators. The reason that (*a*) is not more widely used is the high capital cost of both the equipment and the outside cold 'sink'.

15

It is clear that nothing is to be gained, at any stage, by re-using the rinsings from an earlier stage, since this would only serve to increase the concentration of the residue. If a volume of water $k_n R$ (k_n is a numerical factor to be determined) is added during the nth cycle of operations, the amount of dirt remaining is reduced according to $D_n = D_{n-1}R/(R + k_n R)$. Thus, after p operations,

$$D_p/D_0 = \prod_{n=1}^{p} 1/(1 + k_n).$$

If k_n is kept at a constant value, k, then the number of operations to use all the water is W/kR, and so $D/D_0 = (1 + k)^{-W/kR}$, i.e.

$$\ln\left(\frac{D}{D_0}\right) = -\frac{W}{R}\frac{\ln(1 + k)}{k}.$$

The largest value of the right-hand side ($= W/R$) occurs when $k = 0$, and thus the smallest value of D is given by $D = D_0 \exp(-W/R)$ and is obtained by using as little water as possible for each operation. It is clear from this result that no improvement results from trying any procedure other than keeping k_n constant.

16

If the external field is denoted by H, then the specimen will develop a dipole moment which is proportional to H. If it is treated as a small sphere of radius r, the dipole moment is given by

$$m = 4\pi r^3\left(\frac{\mu - 1}{\mu + 2}\right)H \simeq \frac{4\pi r^3 \chi H}{3} = V\chi H$$

where V is the volume. This will, in turn, produce a field at the coil, its value on the axis being $2\mu_0 m/4\pi x^3 = B$, say, where x is the distance between coil and specimen. If the dimensions of the coil are small compared with x, the field may be taken to be constant over the whole area, and thus the flux is $\Phi = AB = 2\mu_0 AV\chi H/4\pi x^3$. The motion of the coil can be described by $x = x_0 + a \sin \omega t$, and the EMF per turn of the coil is $d\Phi/dt$, which is easily found to be

$$\frac{3\mu_0 AV\chi H\omega a \cos \omega t}{2\pi x_0^4}.$$

17

A beam of electrons, moving with velocity v and carrying a current I, produces a magnetic field H which, at a distance r from the centre of the beam, has a magnitude $H = I/2\pi r$ and a direction along the circumference of the circle of radius r. An electron on the periphery of the beam, charge e, will experience a force $e\boldsymbol{v} \times \boldsymbol{B}$ ($\boldsymbol{B} \equiv \mu\mu_0 \boldsymbol{H}$) directed radially inwards. Its magnitude is thus $ev\mu\mu_0 I/2\pi r$.

The beam has a line charge density $\rho = I/v$ and the same electron will therefore also experience a Coulomb force of repulsion, i.e. directed radially outwards, of magnitude $\rho/2\pi\epsilon\epsilon_0 r$, i.e. $Ie/2\pi\epsilon\epsilon_0 rv$. The ratio of the electric to the magnetic force is thus $1/v^2\epsilon\epsilon_0\mu\mu_0$. In a vacuum, $\epsilon = \mu = 1$. Also, $\epsilon_0\mu_0 = 1/c^2$. Thus the ratio is c^2/v^2, which is greater than unity, i.e. the beam will defocus.

An alternative treatment is to consider the motion in a frame of reference which is moving along with the electrons. There are then no first-order magnetic fields, but the electrostatic repulsion remains. On transforming back to the laboratory frame, this repulsion still gives a divergent motion, albeit at a reduced speed.

In a plasma, the situation is very much more complicated. It is possible for an excess of positive ions to persist along the path of the electron beam, the negative ions having diffused away rather more rapidly, and this will give rise to a focusing effect. But this is outside the scope of the discussion prompted by the quotation given.

18

(*a*) A telephoto lens, like a telescope, makes far objects appear near by giving rise to an angular magnification. Let us approximate a horse by a rectangular body with a vertical leg at each corner. Suppose the separation of the forelegs, and of the back legs, is 0.5 m, and the distance between front and back legs is 2.0 m. If we observe the horse, head-on, when it is 10 m away, the gap between its forelegs will subtend an angle of 1/20 at the eye, while the gap between the hind legs subtends 1/24. It is this difference which gives the impression of the length of the horse. If the same horse is 100 m away the same two angles become 1/200 and 1/204, which are not very different. However, if we use a telescope with an angular magnification of ×10, the angles become 1/20 and 1/20.4. The apparent size of the front is the same as that of a horse standing 10 m away, but now its hind legs appear to be only 0.2 m further back. A very short horse!

(*b*) In order to reduce image defects such as astigmatism and field curvature in a camera lens, an aperture stop is often placed on the optic axis a little way in front of the lens, as shown in the diagram. The result is that images of off-axis points are formed by a restricted area of the lens, which is itself off axis. The image formed by a narrow cone of rays passing through the centre of the lens would be at P. It is clear from the diagram that, in addition to focusing the cone of rays which actually forms the image, the lens must also deflect its principal ray, acting, in effect, like a small-angle prism. The amount of this deflection will be determined by design parameters of the lens already fixed, and the off-axis pencil forms its image at P′, closer to the axis than P. It is also fairly clear that, the larger the object, the more eccentric will be the patch of lens used to form the image, and thus the larger the error PP′. The magnification is thus progressively reduced as we move further away from the axis. This is barrel distortion. It follows that the effect will be particularly noticeable with a wide-angle lens.

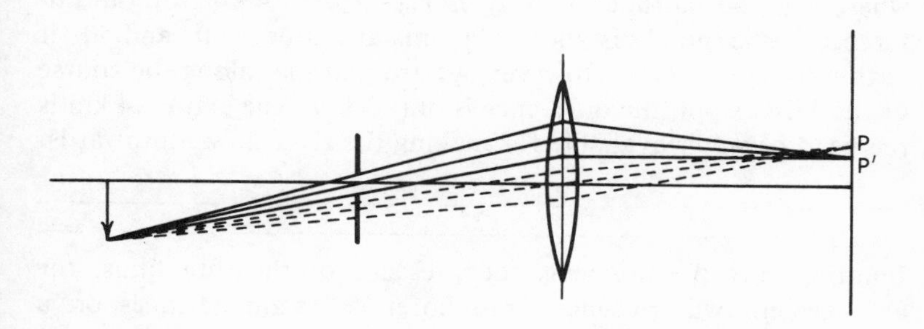

19

Initially we have

$$mv^2/r = Gm^2/4r^2. \tag{1}$$

The final system is most likely to break up if the explosion increases the velocity of one component in the direction in which it is moving at the time. Call the increased velocity v_f. The centre of mass now has a velocity $(v - v_f)/2$, and the velocity of either star referred to axes in which the centre of mass is stationary is, say,

$$(v + v_f)/2 = v'. \tag{2}$$

The system will break up if the new kinetic energy is greater than the work needed to separate the components to infinity, i.e. if $mv'^2 > Gm^2/2r$. Using the relations (1) and (2), the required condition becomes

$$v_f = (2\sqrt{2} - 1)v = (2\sqrt{2} - 1)(Gm/4r)^{1/2}.$$

The above argument assumes that the mass lost as a result of the explosion is so small that it has no effect on the result. If this quantity is denoted by δm, then an analysis on exactly the same lines gives the condition $v_f/v = 2(2 - \delta m/m)^{1/2} - 1$, thus showing that the assumption was valid.

20

The earth is not a sphere, but is, approximately, an oblate spheroid, with its equatorial diameter a about 21 km bigger than its polar diameter b. The equation for an ellipse can be written as $x = a \cos\theta$, $y = b \sin\theta$. Writing $r^2 = x^2 + y^2$ and $a - b = \delta$, and neglecting squares of δ/a, this can be written as $r = a - \delta \sin^2\theta$ where θ is the latitude. This gives $r_{30} - r_{50} = 7.4$ km for the difference in the radial distances of points at latitudes 30° and 50° if both were at sea level. However, we are told that along the course of the Mississippi, the difference is only 5 km. The extra 2.4 km is the head of water available for making the river flow southwards.

21

Ignoring, for the moment, the presence of the thin films, the arrangement will produce Fraunhöfer diffraction fringes on a screen in the focal plane. The intensity in terms of the angle of diffraction θ is given by the well known result

$$I \propto \left(\frac{2a \sin\phi}{\phi}\right)^2$$

where $\phi = 2\pi a \sin\theta/\lambda$ and a is the slit width. The first zero on either side of the bright central fringe is at $2\pi a \sin\theta/\lambda = \pi$ or, in terms of distances on the screen, at $x = \pm\lambda f/2a$, where f is the focal length of the lens. If white light is used, then the presence of λ in this expression shows that little more than the central fringe will be observed.

However, the thin films described act as an interference filter. Repeated internal reflections inside the silica film (thickness d, refractive index μ) will give rise to constructive interference only if $2\mu d = n\lambda$ with $n = 1, 2, 3 \ldots.$ For the values specified, $\lambda = 12\,000$, 6000, 4000, 3000 ... Å. The first and last wavelengths are in the infrared and ultraviolet, respectively, so that the filter will transmit 6000 Å (yellow) and 4000 Å (blue).

The effect observed will depend on the meaning of the phrase 'half-reflecting'. If this is taken literally to mean a reflection coefficient of 0.5, then intermediate wavelengths will also be transmitted with appreciable intensity and the result will not be very different from that observed for white light. The ratio of maximum intensity at the wavelengths specified to minimum intensity in between is equal to $(1+R)^2/(1-R)^2$, where R is the reflection coefficient. (This is another standard result, but is not difficult to prove by summing the infinite series of vibrations whose amplitudes decrease in geometrical progression while their phases increase in arithmetical progression.) For $R = 0.5$, the ratio is $9:1$ and the transmitted band of wavelengths is very broad. But if R is increased to 0.9, the ratio becomes $360:1$ and, moreover, the transmitted waveband is quite narrow. There will thus be two diffraction patterns in the visible part of the spectrum, one blue and one yellow, their central maxima being coincident. The first zero is thus at $x = 0.67$ mm on the screen for the blue fringes but at $x = 1.0$ mm for the yellow set, and a few coloured fringes will be seen before they merge. Unfortunately, in order to obtain high reflection coefficients with silver, the films must be made thicker, so that the absorption increases and the fringes will be faint.

22

The ground-state wavefunction for hydrogen is

$$\psi(r) = (\pi a^3)^{-1/2} \exp(-r/a)$$

where the Bohr radius is

$$a = \frac{\hbar^2}{m} \frac{4\pi\epsilon_0}{e^2}.$$

The potential energy is

$$V(r) = -\frac{e^2}{4\pi\epsilon_0} \frac{1}{r}$$

and the ground-state energy is

$$E = -\frac{1}{2} \frac{e^2}{4\pi\epsilon_0} \frac{1}{a}.$$

Thus for $r > 2a$ the value of $V(r)$ exceeds E. An observation of the proton–electron spacing r may result in any value, but the act of measurement of r disturbs the system. Since r does not commute with the Hamiltonian, the radius and energy cannot simultaneously be measured. Enough energy will be put into the system to raise the energy above the potential energy at any point where the electron may actually be observed.

23

Let a, b denote the inner and outer radii of the annular space between the rod and the cylinder and V the potential of the rod relative to the cylinder. Then the field $F = V/r\ln(b/a)$. If the polarisability of a particle is α, then the force on it is $\alpha F \partial F/\partial r$. The equation describing its radial motion is thus

$$m \frac{\mathrm{d}^2 r}{\mathrm{d}t^2} = \frac{\alpha V^2}{r^3 [\ln(b/a)]^2}$$

the solution of which is

$$t = \frac{r_0(r_0^2 - r^2)^{1/2}}{V} \left(\frac{m}{\alpha}\right)^{1/2} \ln\left(\frac{b}{a}\right).$$

m is the mass of the particle, and the initial conditions $r = r_0$, $\mathrm{d}r/\mathrm{d}t = 0$, at $t = 0$ have been applied. If we now set $r_0 = b, r = a$, this gives τ, the time needed for all the particles to reach the central electrode. If the flow velocity is v, then the axial distance travelled in this time is $v\tau$ and, if all the particles are to be collected, this must equal L, the length of the cylinder. Hence we obtain

$$\frac{L}{v}=\frac{b(b^2-a^2)^{1/2}}{V}\left(\frac{m}{\alpha}\right)^{1/2}\ln\left(\frac{b}{a}\right).$$

24

For any spacecraft in a circular orbit, $GMm/r^2 = mr\omega^2$. For a body on the surface of the earth, $GM_E/r_E^2 = g_E \simeq 10\ \text{m s}^{-2}$. With the numbers given, we find $\omega = 10.4\ \text{rad s}^{-1}$, i.e. the period is approximately $\pi/5$ seconds—somewhat uncomfortable for the intrepid astronaut! We can also deduce that the linear velocity in orbit is approximately $10^7\ \text{m s}^{-1} \simeq 0.03c$—high, but not relativistic.

For a person at the mass centre of the craft, $GMm/r^2 = mr\omega^2$ also, i.e. she will, indeed, float. If r changes to $r-x$, the gravitational force becomes $GMm/(r-x)^2$. If sideways motion across the vehicle is prevented, and if the vehicle mass is large compared with the mass of the person, then ω will remain constant, and the force needed to maintain the astronaut in the new, smaller, orbit will be $m(r-x)^2\omega$. This is no longer equal to the gravitational force. The difference, approximately $3mx\omega^2$, will accelerate the person towards the centre at a rate $3\omega^2 x$. This will give rise to a velocity $\sqrt{3}\omega x$. At the end of the tunnel ($x = 50$ m), this is equal to $900\ \text{m s}^{-1}$ (about 2000 mph in more homely units). On the way to this unfortunate conclusion there will be a Coriolis acceleration sideways across the tunnel equal to $2\sqrt{3}\omega^2 x$. This will reach the value g_E before the victim has moved more than a few centimetres.

The orbit as described is indeed stable. The differential radial force calculated above—the so-called tide-raising force—will provide a restoring couple if the attitude is disturbed, as can be seen clearly by drawing a diagram.

25

The height is readily found by simple geometry; it is, to a good approximation, $h = d^2/2r$, where d is the range (500 km) and r is the radius of the earth (6400 km). The height is 19.5 km.

A crucial question is whether the cable can support its own

weight. This requires the tensile strength to be greater than ρgh. Estimating the density of steel at 8×10^3 kg m^{-3}, and that of kevlar at 1.2×10^3 kg m^{-3}, we obtain $\rho gh = 1.59 \times 10^9$ N m^{-2} and 0.28×10^9 N m^{-2} respectively. Thus steel is barely strong enough, while kevlar has a comfortable factor of safety ($\simeq 7$). Some improvement could be obtained by making the diameter of the cable taper downwards.

The design problem involves a number of considerations interrelated in a complicated way. The total load to be supported = equipment + cable + fabric of balloon: only the equipment is a fixed quantity. Horizontal forces due to wind drag on both balloon and cable are important. Their effect would be to make both the position and—more importantly—the height of the transmitter very variable. Their effect could be reduced by making the total buoyancy much more than sufficient to support the load. The buoyancy, the drag force and the weight of fabric all depend on the balloon size. The cable diameter will depend mainly on the excess buoyancy, i.e. on the drag forces. The total buoyancy will depend on the relative densities of air and helium at 20 km altitude. The temperature of the balloon, different by day and by night, will affect both the buoyancy and the mechanical properties of the fabric. Power supply to the transmitter would present a problem, and the hazard to aircraft should not be overlooked.

26

When displaced a distance x towards one of the plates, the sphere finds itself in a position where the potential is Vx/b. It therefore acquires an induced charge, sufficient to keep its potential at zero (it is earthed). Its capacitance is $4\pi\epsilon_0 a$, and so the charge is $4\pi\epsilon_0 aVx/b$. Being in a field V/b, it experiences a horizontal force $4\pi\epsilon_0 aV^2x/b^2$ towards the nearer plate. The tension in the wire has a horizontal component equal to mgx/l, and the system becomes unstable when these are equal. This gives $V^2 = mgb^2/4\pi\epsilon_0 al$.

(There will also be an induced dipole on the sphere, of magnitude $4\pi\epsilon_0 Ea^3$, where $E = V/b$, the field strength. This will interact with the image dipoles in the two plates to produce a net force towards the nearer one. Its magnitude can be shown to be $3(4\pi\epsilon_0)a^6V^2x/b^7$, which is smaller than the monopole interaction in the ratio $(a/b)^5$ and is thus quite negligible.)

27

(i) The radius of the focal spot is $x = 0.61 f\lambda/a$, where f is the focal length of the lens and a is the radius of the laser beam. This gives $x = 1.8 \times 10^{-6}$ m. If we are to neglect conduction, the input energy must be reradiated from a spot of this size. The energy radiated from an area A at temperature T is $\epsilon\sigma A T^4$, where ϵ is the emissivity. Setting $\epsilon = 1$, the resulting value of T is approximately 6.4×10^3 K.

(ii) Consider a spherical heat source, radius r_1, temperature T_1, in an infinite medium with $T = T_\infty$ at infinity. In the steady state, the heat flow through *any* spherical shell concentric with the source is given by $Q = -\varkappa 4\pi r^2\, \mathrm{d}T/\mathrm{d}r$. Integrating, and using the boundary condition at r_1, we obtain $Q = 4\pi\varkappa r_1(T_1 - T_\infty)$. If the piece of lead is treated as a semi-infinite solid, with the source in its plane face, Q will be about half this quantity.

The radiation loss is approximately $\pi r_1^2 \sigma (T_1^4 - T_\infty^4)$, allowing for incident radiation from the surroundings at T_∞. Since, in this case, $T_1 - T_\infty$ will be much smaller than either T_1 or T_∞ (on the absolute scale), we can write this approximately as $\pi r_1^2 \sigma (T_1 - T_\infty) 4T^3$. The ratio of conduction loss to radiation loss is thus $\varkappa/\sigma r_1 T^3$, and this is of the order of 10^7. The radiation loss can thus be ignored. Setting $2\pi\varkappa r(T_1 - T_\infty) = 10^{-3}$ W, we have $T_1 - T_\infty \simeq 2.9$ K.

28

The increased pressure below the wire will lower the freezing point, so that some ice at 0°C will melt. The melt water will flow round to the top of the wire and refreeze. The basic assumption is that the process that determines the rate of this sequence of events is the conduction of heat across the wire. The thermodynamic relation quoted serves to determine δT, if $\delta p = Mg/A$ where A is the area under pressure ($= 0.2 \times 10^{-3}\ \mathrm{m}^2$ in this case). The rate of heat transfer by conduction is $\varkappa A \delta T/t$ ($t = 10^{-3}$ m is the vertical thickness of metal). The rate of heat production is given by the amount of ice melted per second, multiplied by the latent heat, i.e. $L\rho A V$(ρ is the density of ice, V is the velocity of descent). In the steady state, the rate of heat production will equal the rate of heat transfer. Putting these relations together, we obtain

$$V = \frac{\varkappa M g T}{t L^2 A} \frac{\Delta\rho}{\rho^3}$$

where the specific volumes v have been replaced by $1/\rho$ and the approximation that $\Delta\rho \ll \rho$ has been made. The numerical result is $6.4 \times 10^{-6}\ \mathrm{m\,s^{-1}} \simeq 23$ mm/hour.

(The above simple analysis represents the accepted doctrine on this problem. Unfortunately, it appears to be wrong. A more detailed analysis, made since the question was set in the 1964 examination, has shown that the thermal conductivity of the ice and the thermal resistance of the film of water cannot always be neglected. On the other hand, the heat transported by the moving water, the effect of the finite viscosity of the water and the heat generated by the work done all give rise to very small effects. But even when experimental conditions are chosen so that heat conduction across the wire is, indeed, the most important process in determining the speed of descent, experimental results can still be up to ten times smaller than predicted by the appropriate theory.)

29

The force between the magnet and the superconductor can be treated by the method of images, familiar in other contexts. The surface of the superconductor is treated as a mirror and, if the image has the same polarity as the object, considerations of symmetry show that there will be no component of field normal to the surface; that is, the field of the real magnet does not penetrate into the superconductor, as required. Thus the forces between the magnet and the superconductor are the same as those between the magnet and its image. They can be characterised by an interaction energy which can readily be calculated from first principles—e.g. by summing the pair-wise interactions of the four poles. If p is the dipole moment and h is the height of the magnet above the surface, the energy is $(\mu_0/4\pi)[p^2/(2h)^3]$. There is also a gravitational energy mgh, where m is the mass of the magnet. The equilibrium position is obtained by minimising the total energy, i.e. by setting

$$\frac{\mathrm{d}}{\mathrm{d}h}\left(mgh + \frac{\mu_0}{4\pi}\frac{p^2}{(2h)^3}\right) = 0.$$

This gives $h = (3\mu_0 p^2/32\pi mg)^{1/4}$. The result is 1.06 cm.

The above tacitly assumes that the equilibrium position is with the dipole axis parallel to the surface. This is fairly clear from symmetry considerations, but a complete treatment gives

$$\frac{\mu_0}{4\pi}\frac{d^2(1+\sin^2\theta)}{(2h)^3}$$

for the energy, where θ is the angle between the dipole axis and the surface. This is clearly a minimum when $\theta = 0$.

30

This question is not sufficiently precisely formulated to permit an unambiguous answer to be given. For the simplest possible arrangement, we may envisage a large hollow conducting sphere, at earth potential, with a hole at one point, through which the beam of electrons enters, and another, diametrically opposite, through which it leaves. The 10 cm sphere is placed at the centre. The electron beam must be presumed to originate at a cathode which is held at a potential of -1000 V relative to earth. Under these conditions, charge will accumulate on the sphere until its potential is equal to that of the electron source, i.e. -1000 V. If the radius of the outer sphere is much greater than 10 cm, the capacitance of the inner sphere is $4\pi\epsilon\epsilon_0 r$ where r is its radius. If $\epsilon = 1$, the final charge on it will thus be $4\pi\epsilon_0 Vr$, with $V = 10^3$ V, i.e. $Q = 1.11\times10^{-8}$ C. The field at its surface will then be $Q/4\pi\epsilon_0 r^2 = 10^4\ \mathrm{V\,m^{-1}}$. The assumption about no loss of charge is quite unrealistic. Any conductor bombarded by 1000 V electrons will emit secondaries, which will be repelled by the charge on the sphere. This charge will build up until the rate of escape of secondaries just balances the rate of arrival of the primaries. The details will depend on the secondary-emission characteristics of the metal, but the outcome is likely to be a final potential much less than 1 kV.

31

Assume that the grain size of the emulsion is so small that it does not affect this problem. Then to locate the edge of a particle of rock with a resolution of 10^{-1} mm, a density of tracks of at least $100\ \mathrm{mm^{-2}}$ will be required. If these reach the emulsion they must have originated in a layer of water not more than 'a few' (=5) microns thick, i.e. a density of disintegrations of at least $2\times10^4\ \mathrm{mm^{-3}}$. Since half of the tracks will go downwards and not reach the emulsion, we must double this density, i.e. to $4\times10^4\ \mathrm{mm^{-3}}$.

The tritium will decay as $n = n_0 \exp(-\alpha t)$, with $\alpha = 6.5 \times 10^{-6}\ \mathrm{h}^{-1}$ to give the specified half-life. The number of disintegrations is then $\mathrm{d}n = \alpha n\ \mathrm{d}t$. With $\mathrm{d}n = 4 \times 10^4$ and $\mathrm{d}t = 1$ h we obtain $n \simeq 6 \times 10^9\ \mathrm{mm}^{-3} = 6 \times 10^{12}\ \mathrm{cm}^{-3}$.

Practical difficulties would include the problem of preparing a sufficiently smooth surface without disturbing the structure under investigation; effects of the finite grain size of the photographic emulsion; the minimum dosage calculated might give a 'spotty' picture, with a lack of resolution; problems of doing all operations at sub-zero temperatures; problems of radioactive contamination.

32

We assume that the concentration of free electrons is small. If this were not the case, the field strength would vary with position, and more information would be needed. The lines of force will then be as in the diagram, which is drawn for $E_1 > E_2$. If the electrons can be said to 'drift', this implies that they collide frequently with the gas atoms and their average paths will thus follow the lines of force. If the direction of the field is such that they move from E_1 to E_2, then clearly only a fraction E_2/E_1 will penetrate the screen, the remainder impinging on the wires. If the sign of the fields is reversed, all of them will pass through. If the gas pressure were greatly reduced, the electrons would not 'drift' but would be accelerated by the fields. The fact that their mass is finite means that they would not be able to follow the curved field lines, but would move on paths more nearly straight. In the first case above, i.e. electrons moving from E_1 to E_2, a greater number would penetrate the screen, and in the second case a smaller number. In high fields, and any vacuum less than perfect, the collision of the fast electrons with residual molecules could produce electron multiplication, and eventually an avalanche. The possibility of the emission of secondary electrons from the wires must also be considered, if the potential

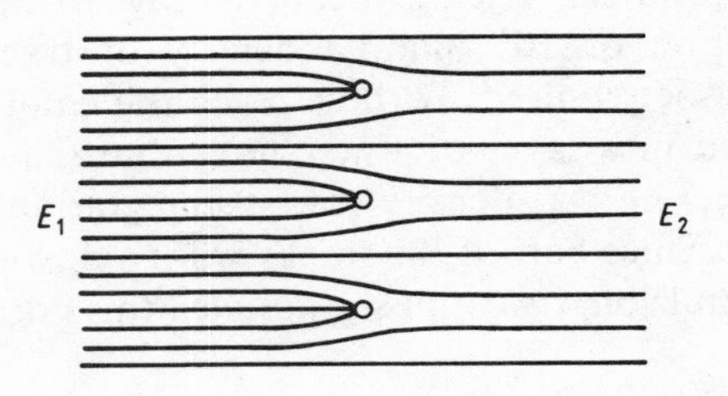

drop along one mean free path is large enough. If the drift is from E_1 to E_2 this would have the effect of increasing the apparent fraction reaching the E_2 region. If the wires are very fine, then the high local field near their surface may give rise to field emission, independent of the impact of electrons.

33

In this application of the uncertainty principle, $\Delta x \Delta p \simeq \hbar$, we can write $\Delta x = \theta_0$ and $\Delta p = I\dot{\theta}_0$, where I is the moment of inertia of the pencil about its tip $(= ml^2/3)$. Thus

$$\theta_0 \dot{\theta}_0 = 3\hbar/ml^2. \qquad (1)$$

Since $\theta(t) = \theta_0 \cosh(t/\tau) + \dot{\theta}_0 \tau \sinh(t/\tau)$, we estimate the greatest time for which it can remain balanced by calculating the minimum value of $\theta(t)$ subject to the condition (1). This is a standard procedure. (The minimum of $f(x, y) \equiv \alpha x + \beta y$, subject to $xy = c$, occurs at $x = (\beta c/\alpha)^{1/2}$ and has the value $2(\alpha\beta c)^{1/2}$.) The result is

$$\theta_m = 2\,[\cosh(t/\tau)\sinh(t/\tau)\,3\hbar\tau/ml^2]^{1/2}$$

which can be rearranged as

$$\sinh(2t/\tau) = ml^2\theta_m^2/6\hbar\tau. \qquad (2)$$

Taking $l \simeq 15$ cm, the period is $\tau \simeq 0.1$ s. With $m \simeq 5$ g and any reasonable value of θ_m, it is clear that the right-hand side of (2) will be very large, and we can therefore write

$$\sinh(2t/\tau) \simeq \tfrac{1}{2}\exp(2t/\tau)$$

and thus

$$2t/\tau = \ln(ml^2\theta_m^2/3\hbar\tau). \qquad (3)$$

If we adopt $\theta_m \simeq 0.01$ rad as the limit at which the pencil can be said to be still balanced, equation (3) gives $t \simeq 3.8$ s. If we take $\theta_m = 0.1$ rad ($\simeq 6°$), this only changes by a small amount, to 4.0 s.

34

The angular frequency is $\omega = (g/l)^{1/2}$. If A is the amplitude, the energy of the oscillation is $E = mA^2\omega^2$ and the tension in the string is $T = mg + m(A\omega \sin \omega t)^2/l$. Since the string is shortened slowly we are justified in calculating the work done by using the time-averaged value of T, which is $\overline{T} = mg + mA^2\omega^2/2l$. The work done in shortening the string by δl is thus $mg\delta l + mA^2\omega^2\delta l/2l$. The first term increases the gravitational potential energy of the bob, while the second term adds to the energy of the oscillation an amount $\delta E = -mA^2\omega^2\delta l/2l$ (the negative sign appears because δl is negative). Thus $\delta E/E = -\delta l/2l$. Since $2\pi f = (g/l)^{1/2}$, we also have $\delta f/f = -\delta l/2l$ and so $\delta E/E = \delta f/f$, which establishes the desired result.

35

Suppose a 'book' contains 60 pages with 300 words per page and 5 letters per word: this gives approximately 10^5 characters. The total number of different characters available, including letters, numbers, stops and spaces is approximately 40. Thus the probability of producing any specified arrangement of 10^5 by chance is $(1/40)^{100\,000} \simeq 10^{-160\,000}$, i.e. $10^{160\,000}$ trials are needed to give an even chance of success.

The age of the universe is $T \simeq 10^{10}$ years $\simeq 10^{17}$s. The size of the universe is $L \simeq cT \simeq 3 \times 10^{25}$ m. The size of a small microchip is $l \geqslant 3 \times 10^{-9}$ m. Thus the number of micros required to fill the universe is approximately $(L/l)^3 \simeq 10^{102}$. The time for a bit instruction is $t \geqslant l/c \simeq 10^{-17}$ s, and for 40 characters we need 6 bits/character. Thus the time required to produce a character is $t' \simeq 6 \times 10^{-17}$ s and the number of characters that a micro could generate is $T/t' \simeq 2 \times 10^{33}$. The number of attempts at a book of 10^5 characters is thus approximately 10^{28} from each micro. The total number from all the micros is thus 10^{130}—which is infinitesimal compared with the requirement of $10^{160\,000}$.

If we invert the question, we could ask what is the maximum number of specified characters that could have been produced. The answer is n, where $40^n \simeq 10^{130}$, i.e. $n \simeq 81$: not a book, but a phrase half as long as Huxley's statement.

Acceptance of some errors would have no effect. If there were 1000 permissible errors in the book, it would mean that only $(10^5 - 1000)$ characters were specified—a negligible change. The second question is unanswerable. By way of illustration—it would take $40^4 \simeq 2.5 \times 10^6$ attempts to have an even chance of producing a specified four-letter word. But the number to produce *any* four-letter word depends on the number of meaningful combinations of four letters that exist (in English?).

36

Let v_y, v_x be the initial transverse and longitudinal velocities respectively. The corresponding accelerations are 0 and $Ee/m = f$. If t is the time of flight, the velocities at the aperture are v_y and $v_x + ft \simeq ft$. At the aperture, the inclination of the trajectory to the axis is $\theta = \tan^{-1} v_y/ft$, and the point of impact is a distance $v_y t$ from the axis. All trajectories, extrapolated backwards, therefore cut the axis at a point a distance X from the aperture, where $v_y t/X = \tan\theta$. These relations give $X = ft^2$. However, the real distance travelled parallel to the axis is $d = v_x t + ft^2/2 \simeq ft^2/2$. Thus $X = 2d$.

The application of an axial magnetic field leaves the axial motion unaltered, but the motion in the yz plane turns into a circle of radius mv_y/Be, which is traversed in a time $2\pi mB/e = \tau$, say. The trajectories thus become helices, which project on to the yz plane as circles, with radii $\propto v_y/B$, all passing through the point (0, 0). The magnitudes of the impact velocities are unchanged, but the plane containing the x and y velocities on impact no longer necessarily contains the x axis also, and so the concept of the 'virtual' image is not valid. However, since τ is independent of v_y, a suitable choice of B ensures that all electrons complete an integral number of cycles in travelling the distance d. Thus a 'real' image is formed at (0, 0) for these values of B.

37

In copper the protons (and the neutrons) are not stationary, but move around with 'Fermi momentum' as a quantum mechanical consequence of their confinement to the nuclear volume. When one of them happens to be approaching the incident proton, the threshold momentum will be reduced. Suppose the Fermi momentum is x (small compared with m, the mass of a nucleon). (All quantities are expressed in the customary units, with $c = 1$.) Then the initial state has one proton with momentum p and total energy E_1 and one nucleon with momentum $-x$ and total energy m. (The kinetic energy will be $x^2/2m$, with x small, and so can be neglected.) The final state has four moving particles, all of about the proton mass m. At the threshold, all are stationary in the centre-of-mass frame and thus all have the same momentum, p_2, and the same energy, E_2, in the laboratory frame. The two conservation laws thus give $p_1 - x = 4p_2$ and $E_1 + m = 4E_2$. We also have the general relations $E_1^2 = p_1^2 + m^2$ and $E_2^2 = p_2^2 + m^2$. By manipulating these equations we can obtain $7m^2 = mE_1 + xp_1 - x^2/2$. Since x is small, the last term can be neglected, and, in view of the relation $E_1^2 = p_1^2 + m_1^2$, we can write $p_1 \simeq E_1$ in the penultimate term. This gives $E_1 = 7m/(1 + x/m)$. For $x = 0$, this becomes $E_1 = 7m$; hence $p_1 = \sqrt{48}\, m = 6.5$ GeV/c, as stated. As x increases, the threshold energy E_1 and thus the threshold momentum p_1 decrease: $p_1 = 4.8$ GeV/c when $x = 0.35m = 0.33$ GeV/c.

The reaction probability for incident protons of small momentum will depend on the chance of meeting a nucleon of high momentum—i.e. on the shape of the 'tail' of the distribution curve. Although nucleons obey Fermi statistics, we can probably use a Boltzmann-type distribution for the tail; thus the reaction probability will fall off rapidly with decreasing incident momentum.

38

Let v denote the velocity of sound, and αv that of the rocket, with $\alpha > 1$. Consider a sound wave emitted from the rocket at $t = 0$, when the rocket is at O in the diagram. It will arrive at a point P, with coordinates r, θ as shown, at a time $t_1 = r/v$. In a time τ, the rocket has moved to O′, with OO′ $= \alpha v\tau$. Another sound wave, emitted at this instant, arrives at P at a time $t_2 = \tau + R/v$, where R is the distance O′P. The time interval between the two arrivals, i.e. $t_2 - t_1 = \Delta$, say, is then $\tau + R/v - r/v$. We also have the geometrical

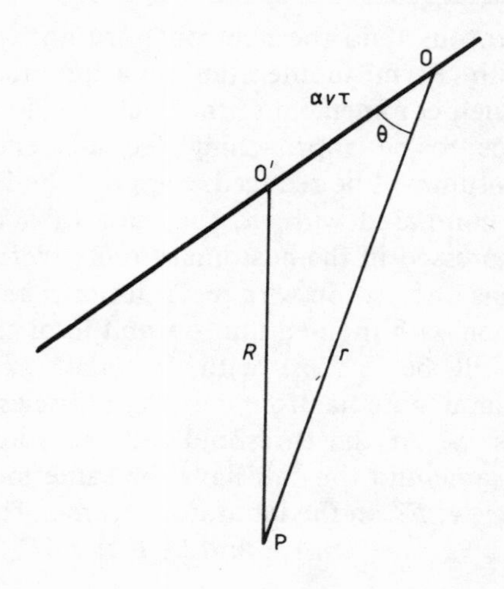

relation $R^2 = r^2 + \alpha^2 v^2 \tau^2 - 2r\alpha v\tau \cos\theta$. Now let us choose θ so that $\cos\theta = 1/\alpha$. Since $\alpha > 1$, this is always possible. The relation can then be written $R^2 - r^2 = \alpha^2 v^2 \tau^2 - 2rv\tau$. For small τ, $R^2 \simeq r^2$ and $R^2 - r^2 \equiv (R-r)(R+r) = 2r(R-r)$. This gives $R - r = \alpha^2 v^2 \tau^2/2r - v\tau$. We can now write $\Delta \equiv \tau + (R-r)/v$ as $\Delta = \alpha^2 v\tau^2/2r$, i.e. Δ is the second order of small quantities in τ. Thus, to a first approximation, all sound waves emitted in the neighbourhood of O will arrive at P at the same time. There will therefore be a shock front at P. This conclusion does not depend on giving any particular value to r, so that all points along OP will experience the shock wave. The locus of points such as P is defined by $\cos\theta = 1/\alpha$. It will thus be a cone, of semi-angle θ, with OO′ as its axis. This will cut the surface of the ground in an ellipse.

39

If the rotor, of radius R, gives a downward velocity v to a column of air initially at rest, the momentum transferred per second will be mv, where m, the mass of the air, is $\pi R^2 v\rho$. If the helicopter is to be airborne, this must equal Mg. Inserting the values given, we obtain $v = 4.5\ \text{m s}^{-1}$. Neglecting energy losses, the energy transferred per second to the bulk motion of the air is $\frac{1}{2} mv^2$, which equals 4.5 kW in this case. For the system to be practicable, the man must thus generate about 5 kW continuously. Making an estimate from the time taken to run, fast, up a long flight of stairs, it appears that the maximum rate of energy production by a man is of the order of 1 kW. The proposal does not, therefore, look very hopeful.

40

The time taken by the particle to fall through the beam is readily found by elementary methods. It has a maximum value of 37×10^{-3} s if the path happens to be along the diameter of the beam.

The electrostatic capacitance of the particle is proportional to its radius (equal in ESU) and is equal to 5.5×10^{-16} F. Hence to charge it to 3 kV we need 1.65×10^{-12} C. The current flowing to it is initially $I(r/R)^2$ (r and R are the radii of the particle and the beam and I is the current in the beam). This is equal to 6.25×10^{-10} A. At this rate, the time of charging would be $1.65 \times 10^{-12}/6.25 \times 10^{-10} = 2.64 \times 10^{-3}$ s. In fact, as the particle accumulates charge, it will repel some of the protons sideways, so that the rate of charging will decrease. However, the charging time of about 2.5×10^{-3} s is so much smaller than the time of fall, 37×10^{-3} s, that, for most trajectories, the particle will be fully charged to 3 kV by the time it leaves the beam.

In addition to charge, the particle will also collect energy and momentum. The energy extracted from the beam is equal to QV. Of this, $\frac{1}{2}QV$ is stored in the electrostatic energy of the charged particle and the other $\frac{1}{2}QV$ appears as heat. This amounts to 2.5×10^{-9} J. Assuming the density of graphite to be about 3×10^3 kg m^{-3}, the resulting temperature rise is 2 K. This will be the average over the particle: locally and temporarily, the rise may be greater.

The velocity of the incident protons is found by equating $\frac{1}{2}mv^2$ to eV, and is found to be about 7.6×10^5 m s^{-1}. Conservation of momentum shows that, if the proton is absorbed, it will impart a velocity of approximately $(m/M)v$ to the particle. Late arrivals will be moving more slowly on impact, but we must then take account of the electrostatic repulsive force between the approaching proton and the partly charged particle. Even when the particle is fully charged, this effect will persist in the quasi-elastic collisions. For a proton approaching head-on, and repelled along the same line, the velocity increment of the particle is $2(m/M)v$. This will fall to zero for off-axis collisions. We may therefore reasonably assume an average velocity increment of $(m/M)v$ for *all* the protons in a beam of cross sectional area πr^2 for the whole of the time t that the particle is in the beam, i.e. about 30 ms for an average trajectory. The resultant velocity is thus

$$\left(\frac{r}{R}\right)^2 \frac{I}{e} t \frac{m}{M} v$$

where e is the proton charge. This gives a value of approximately 20 cm s^{-1}.

41

Let ω be the angular velocity of the axle (a constant) and θ the deflection of the bar from the vertical. Then the relative angular velocity is $\omega - \dot{\theta}$, and the frictional couple is $G = G(\omega - \dot{\theta})$. Since $\omega >> \dot{\theta}$ we can expand this as a Taylor series about ω and write

$$G = G(\omega) - \dot{\theta}(\mathrm{d}G/\mathrm{d}\omega)_{\dot{\theta}=0} + \ldots = G(\omega) - \dot{\theta}\dot{G}(\omega).$$

In equilibrium, with the bar at rest but the axle rotating, we have the relation

$$G(\omega) + 0 = Mgl \sin\theta_0 \tag{1}$$

where M is the mass of the bar and $2l$ is its length. In general,

$$\begin{aligned} I\ddot{\theta} &= G(\omega - \dot{\theta}) - Mgl \sin\theta \\ &\simeq G(\omega) - \dot{G}(\omega)\dot{\theta} - Mgl\theta \end{aligned}$$

or, using (1),

$$I\ddot{\theta} = Mgl(\theta_0 - \theta) - \dot{G}(\omega)\dot{\theta}.$$

Writing $\theta - \theta_0 = \phi$, this becomes

$$I\ddot{\phi} + \dot{G}(\omega)\dot{\phi} + Mgl\phi = 0.$$

If $\dot{G}(\omega) < (4IMgl)^{1/2}$, this represents an oscillation about $\theta = \theta_0$; otherwise, an exponential relaxation towards θ_0. In case (*a*), $\dot{G}(\omega) > 0$ and the amplitude of the oscillations will decrease exponentially with time. In case (*b*), $\dot{G}(\omega) < 0$ and the amplitude will increase exponentially.

42

In the laboratory system of coordinates the initial energies of the two particles are E_0 and mc^2, and their total momentum is $p_0 + 0$, where E_0 and p_0 are related by $E_0^2 = p_0^2c^2 + m^2c^4$. In the centre-of-momentum system, the total energy is E_0' (to be determined) and the total momentum is zero, by definition. Using the concept of invariant mass, we have

$$(E_0 + mc^2)^2 - c^2p_0^2 = (E_0')^2$$

i.e.

$$\begin{aligned} (E_0')^2 &= E_0^2 + 2E_0mc^2 + m^2c^4 - c^2p_0^2 \\ &= 2E_0mc^2 + 2m^2c^4 \end{aligned}$$

using the relation between E_0 and p_0 quoted above.

After the collision, the least energetic configuration will be with three electrons and one positron all at rest in the centre-of-momentum system, i.e. $E_1' = 4mc^2$. The conservation of energy requires that $E_1' = E_0'$, i.e.

$$16m^2c^4 = 2E_0mc^2 + 2m^2c^4$$

which gives $E_0 = 7mc^2$. Thus the kinetic energy of the initial electron must have been $E_0 - mc^2 = 6mc^2$.

43

Case (1). Inside the cavity the field will be uniform, and outside, a dipole field will be added to the uniform applied field. This is a standard result. The potential is $V = -Fz$ inside the cavity and $-Ez + P\cos\theta/r^2$ outside, where E is the applied field and F, P are to be determined ($z \equiv r\cos\theta$). The boundary conditions at the surface of the cavity are (a) equality of the tangential components of the field, which gives $F = E - P/a^3$, and (b) equality of the normal components of displacement, which gives $F = \epsilon E + 2\epsilon P/a^3$. From these two relations we deduce $F = 3\epsilon E/(1+2\epsilon)$, which is equal to $1.23E$ in this particular example. Breakdown will occur when $F = 100\ \mathrm{kV\,cm^{-1}}$, i.e. when $E = 81.3\ \mathrm{kV\,cm^{-1}}$ or when the applied voltage is 40.6 kV. After breakdown, ions from the discharge will accumulate on the surface of the cavity until the field inside falls below the extinction value for the discharge.

Case (2). If the polyethylene has a small conductivity, the initial field distribution (as above) will cause currents to flow, building up a charge distribution on the surface of the cavity. Since there can be no current flow normal to the surface, this process will cease when the radial component of field outside $r = a$ has fallen to zero. The boundary conditions are now (a) continuity of the tangential field, giving $F = E - P/a^3$ as before, and (b) zero radial field for $r \geqslant a$. The second condition gives $E + 2P/a^3 = 0$, and the two together give $F = 3E/2$. The applied voltage for breakdown is now 33.3 kV.

If a voltage between 33.3 and 40.6 kV is suddenly applied, there will be no immediate breakdown but, after some delay, the situation will be as in Case (2), and breakdown will occur. The delay will be of the order of the relaxation time for the dielectric. For a notional parallel-plate capacitor, $C = \epsilon\epsilon_0 A/d$ and $R = \rho d/A$, where ρ is the resistivity, so that $\tau \equiv RC = \epsilon\epsilon_0\rho$, which, using the data given, is approximately 200 s.

44

'Static friction' implies a tangential force at the point of contact of the two solids. The essence of rolling motion is that there is no (macroscopic) relative motion of the two bodies at this point.

Hence the forces do no work, no energy is dissipated, and the wheel is not slowed down.

A more realistic treatment takes account of the fact that both wheel and road will be deformed, to give a finite area of contact. The forces acting across this (curved) area will have components which are not vertical, and will not be symmetrical in a fore-and-aft direction. There may thus be a resultant horizontal component opposing the motion.

A more useful approach is in terms of energy. The deformations of the two bodies due to the high local stresses over the small area of contact may differ in such a way that they give rise to localised, microscopic slipping, resulting in energy dissipation. In addition, the high local stresses will give rise to large local strains—too large to be perfectly elastic. The fact that the behaviour is non-linear would not, in itself, be of any consequence, but such non-linear behaviour is usually accompanied by hysteresis effects, so that some net work is done in a cycle of strain and, again, energy is dissipated.

Even in the limit of very hard solids, very lightly loaded (a small ball-bearing on a glass plate, for example), where the strains may be almost perfectly elastic, they will still be non-uniform. Strain gradients are, as a thermodynamic necessity, accompanied by temperature gradients, and the resulting conduction of heat involves a dissipation of energy ('thermo-elastic damping').

45

If we make the assumption that the optimum trajectory is a parabola, with an initial slope of 45°, then arguments from elementary mechanics show that the initial horizontal and vertical components of the velocity of the shot are each about 90 $\mathrm{m\,s^{-1}}$. The horizontal recoil velocity of the cannon will thus be 4.5 $\mathrm{m\,s^{-1}}$. If the restraining mechanism brings it to rest in, say, 3 m, its deceleration will be 3.4 $\mathrm{m\,s^{-1}}$ and the corresponding force will be approximately 10^3 N—which is not impossible.

In order to discuss the vertical recoil in the same way we would need some estimate of the effective mass of those parts of the ship which move. It is easier to argue as follows. If we assume that the length of the barrel is, say, 3 m, and that a constant force acts on the shot while it is moving up the barrel, then its vertical component of acceleration is $3 \times 10^3\ \mathrm{m\,s^{-2}}$, which gives a vertical force of

3×10^4 N. This will also be the recoil force acting on the cannon + ship. This also should be tolerable.

If, as is likely, the trajectory is more nearly horizontal than was assumed above, then the initial velocity would have to be larger, and the horizontal component would be larger still. This might give trouble with the horizontal recoil.

46

The energy, W, is $\int E^2 \, \mathrm{d}V/8\pi$ taken over the whole of space (except for a sphere of radius $b \to 0$, around $+q$). Thus

$$8\pi W = \int_{\mathrm{P}} E^2 \, \mathrm{d}V + \int_{\mathrm{Q}} E^2 \, \mathrm{d}V$$

where P, Q are the volumes inside and outside the sphere, respectively.

Let us write $\boldsymbol{E} = \boldsymbol{E}_1 + \boldsymbol{E}_2$, where $\boldsymbol{E}_1$ is due to the point charge $+q$ and $\boldsymbol{E}_2$ is due to the distribution $-q$. Then, within P, $\boldsymbol{E}_1 = q/s^2$ and $\boldsymbol{E}_2 = 0$, where s is the distance of element $\mathrm{d}V$ from the charge $+q$. Within Q, $\boldsymbol{E}_1 = q/s^2$ as before, and $\boldsymbol{E}_2 = q/t^2$, where t is the distance of $\mathrm{d}V$ from the centre of the sphere.

Now, $E^2 = (\boldsymbol{E}_1 + \boldsymbol{E}_2) \cdot (\boldsymbol{E}_1 + \boldsymbol{E}_2) = E_1^2 + E_2^2 + 2E_1E_2 \cos\theta$. From symmetry, for every volume element where $\cos\theta = a$, there will be another similar volume element where $\cos\theta = -a$. Thus the integral of the third term over space is zero, and so

$$8\pi W = \int_{\mathrm{P}} (E_1^2 + E_2^2) \, \mathrm{d}V + \int_{\mathrm{Q}} (E_1^2 + E_2^2) \, \mathrm{d}V$$

$$= \int_{\mathrm{P+Q}} E_1^2 \, \mathrm{d}V + \int_{\mathrm{Q}} E_2^2 \, \mathrm{d}V$$

which is independent of the position of $+q$. Thus the energy of the 'external' field which appears in Q is compensated by a decrease in the 'internal' field in P, as $+q$ moves away from the centre.

47

Mechanisms of energy dissipation are of two kinds. (1) Those concerned with the vehicle itself, which absorb a finite amount of energy. These are the heating, melting and evaporation of the structure. The heat shield is present to confine these effects to itself, and to protect the rest of the vehicle. Calculations show that the total amount of energy that can be absorbed in this way is only a small fraction of the kinetic energy. We must therefore rely mainly on (2), those involving the interaction of the vehicle with its surroundings. These include radiation and interaction with the atmosphere. The former is significant, but the latter is the dominant process. It consists essentially of collisions with air molecules. Most of the energy involved in these collisions is given to the air molecules: the collisions are 'elastic'. However, some will be inelastic and the energy will remain in the heat shield, to give the effects discussed under (1). The shield should therefore be designed to carry out this function while protecting the rest of the vehicle. The fact that it absorbs some energy is useful, but somewhat incidental: the crucial point is that it should not melt or disintegrate. The principal design requirements for the heat shield are thus

(*a*) a high melting point, and a high latent heat so that when it gets hot—as it will—it does not melt;

(*b*) a low thermal diffusivity, i.e. low thermal conductivity, high density and high specific heat, so that the heating effects are confined as much as possible to the outer layers;

(*c*) as large a thickness as possible, for the same reasons of thermal insulation. Increases in density and in thickness will both increase the mass and there will be an upper limit on what is allowable, for other reasons;

(*d*) high thermal and mechanical strength, to enable it to resist the internal stresses set up by the differential heating without mechanical failure.

(Order-of-magnitude quantitative arguments in support of the above contentions can be given thus. We need, as a preliminary, estimates of several quantities. (i) The velocity: if the orbit radius is equal to the earth radius, then the velocity is given by $mv^2/r = mg$: hence $v \simeq 10^4$ m s^{-1}. (ii) The size: say a sphere of radius 2 m. (iii) The mass: about 10^4 kg—at least, this gives a mean density of about 300 kg m^{-3}, i.e. the vehicle floats fairly high in the water, as observed. (iv) The ratio of the mass of the heat shield to the total mass, m/M. The designer would keep this as small as possible, say 1/10. For mechanisms under (1) above, the energy to be disposed of is 5×10^7 J kg^{-1} referred to the whole vehicle, which is 5×10^8 J kg^{-1} of the heat shield. Most materials have specific heats that are much less than that of water, and materials with high melting points have latent heats that are a good deal larger than that of water. Taking the high temperature as 2000 K, this suggests about 8×10^5 J kg^{-1} from the specific heats and about 2×10^6 J kg^{-1} from the latent heats—say 3×10^6 J kg^{-1} together. This is quite inadequate, even with allowances for errors of estimation, to match the 5×10^8 J kg^{-1} required.

For calculations under heading (2), we need an estimate of the re-entry time. Considerations of the physiologically tolerable values of acceleration are not relevant, since they predict only a minimum time, which is much too short for our purposes. For purposes of estimation, let us try 10^3 s. The necessary rate of energy dissipation is thus 5×10^8 W. With a surface area for radiation of approximately 20 m^2 and a temperature of 2000 K, the rate of radiation loss is about 2×10^7 W—appreciable, but not enough.

For an elastic head-on collision between the vehicle of mass M and a molecule of mass μ, the former with a velocity V and the latter at rest, an application of the conservation of energy and momentum leads to the conclusion that the kinetic energy of M will be reduced by $2\mu V^2$. For an average collision, we can halve this figure. The loss of energy per second is thus $\mu V^2 nAV = \rho AV^3$, where n is the number density of molecules and A is the projected area of the vehicle. Now ρ at ground level is $\rho_0 \simeq 1$ kg m^{-3}, but decreases with height according to $\rho = \rho_0 \exp(-h/h_0)$, where h_0, the scale height, is $k_B T/\mu g \simeq 10^4$ m. If the vehicle starts at a height of 100 km ($= 10h_0$), $\rho = \rho_0 e^{-10} = 10^{-4}\rho_0$. The energy loss rate is thus 10^9 W. This is more than sufficient, even allowing for errors of estimation. As the height decreases, ρ will increase, but V^3 will decrease rapidly. The net effect will be complicated. However, it may be possible to reduce the descent time below the estimated 10^3 s. This is desirable, since it gives less time for heat to penetrate to the inside of the heat shield.)

48

Symmetry suggests that the hole will remain circular, and we assume that the film ahead of the advancing edge is undisturbed. Then, considering an angular segment $\delta\theta$, the accumulated mass of liquid at the edge of the hole is $\frac{1}{2}\sigma\delta\theta r^2$. The outward force acting on this element is $2Tr\delta\theta$. Since $F = d(mv)/dt$, we get $d(r^2 v)/dt = 4Tr/\sigma = 2v_0^2 r$, say, where $v = dr/dt$ and $v_0^2 = 2T/\sigma$. This gives $r\, dv/dt + 2v^2 = 2v_0^2$. If we assume that $dv/dt = 0$, apart from an initial transient, we see at once that $v = v_0 = (2T/\sigma)^{1/2}$. If we write $dv/dt = v\,dv/dr$, we can integrate the equation exactly to give $v^2 = v_0^2(1 - r_0^4/r^4)$, where r_0 is defined by $v = 0$ at $r = r_0$. This justifies the assumption made.

Note that an argument based on energy conservation gives the wrong result. If we equate the kinetic energy of the advancing edge to the surface energy of the film that has vanished, we obtain $\frac{1}{2}\sigma A v^2 = 2\gamma A$, where γ is the surface energy, i.e. $v = (4\gamma/\sigma)^{1/2} = (4T/\sigma)^{1/2}$ if we neglect the difference between surface tension and surface energy. The error arises because the impact of the advancing filament of liquid with the undisturbed film is not a conservative process. Some energy will appear as turbulent motion in the liquid.

The surface tension of soap solutions is of the order of $3 \times 10^{-2}\,\mathrm{N\,m^{-1}}$. The thickness of a soap film is a few wavelengths of visible light ($\simeq 10^{-6}$ m), and its density is that of water. We thus obtain about $10\,\mathrm{m\,s^{-1}}$ for the velocity.

49

We choose the positive sense of ω so that the velocity of the point of contact, v_c, is given by $u + a\omega$. The frictional force $F = \mu mg$ is assumed constant. The motion of the centre is given by $u = u_0 - \mu gt$, and the rotation about the centre by $\omega = \omega_0 - a\mu mgt/I$. Since $I = 2ma^2/5$, the latter expression becomes $\omega = \omega_0 - 5\mu gt/2a$. These combine to give $v_c = u_0 + a\omega_0 - 7\mu gt/2$. This is equal to zero (i.e. slipping stops and rolling begins) when $t = 2(u_0 + a\omega_0)/7\mu g = \tau$, say. The value of u at $t = \tau$ is $a\omega_0(5u_0/a\omega_0 - 2)/7$; this gives the speed of rolling after slipping has stopped. If $u_0/a\omega_0 < \frac{2}{5}$, the value is negative, and the ball will roll backwards towards the cue. This corresponds to the imposition of 'backspin' as specified. If $u_0/a\omega_0 > \frac{2}{5}$, the ball will roll forwards.

To enquire whether this is possible, we consider an impulse Q given to the ball at a point $a - x$ above the table, and in a direction making an angle θ with the horizontal (see the diagram). Then $mu_0 = Q\cos\theta$ and $I\omega_0 = Qp$, with $p = a\sin(\theta + \phi)$ and $\sin\phi = x/a$. If we use the relation $5u_0 = 2a\omega_0$ to determine the boundary condition between the two modes of behaviour, then, after a little algebra, we get $\cos\theta = \sin(\theta + \phi)$, i.e.

$$\tan\theta = \left(\frac{1 - x/a}{1 + x/a}\right)^{1/2}.$$

Real values of θ can be found for any x/a—except that $x \to a$ would be a little difficult, in practice!

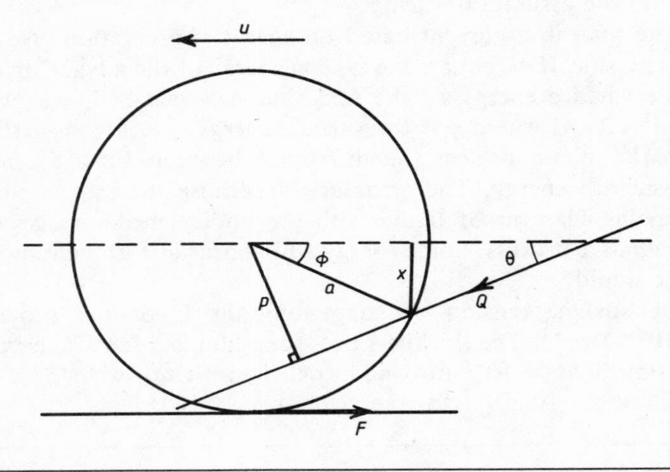

50

The temperature distribution in the wire, in the steady state, is given by the balance between the heat generated by the passage of the current and heat losses (a) by conduction to the ends, (b) by radiation to the surroundings and (c) by conduction and convection through the gas. Approximately,

$$\frac{I^2\rho}{\pi a^2} = \frac{\partial^2\theta}{\partial x^2} k\pi a^2 + 2\pi a[\sigma\epsilon(\theta^4 - \theta_0^4) + c(\theta - \theta_0)]$$

where x is the position, θ is the temperature, ρ is the resistivity, a is the radius, k is the thermal conductivity, ϵ is the emissivity, σ is Stefan's constant and c is a coefficient giving heat losses through the gas. The boundary conditions are $\theta = \theta_0$ at $x = \pm l$, $\partial\theta/\partial x = 0$ at $x = 0$; ρ will depend on θ. A full solution is not practicable. If the wire is long enough, then the high value of the melting point of copper will ensure that near the middle—the hottest point—the loss by radiation will dominate all the others when the fuse is on the point of melting. Thus

$$I_c^2\rho/\pi a^2 = 2\pi a\epsilon\sigma(\theta_c^4 - \theta_0^4).$$

If ρ is a known function of θ, this gives I_c in terms of θ_c, the melting point of copper.

If the current is suddenly switched on, the temperature near the mid-point will rise according to

$$\pi a^2 ds \frac{\partial\theta}{\partial t} = \frac{I^2\rho}{\pi a^2} - 2\pi a\epsilon\sigma(\theta^4 - \theta_0^4)$$

where d is the density and s is the specific heat. This incorporates the same assumptions as before—that there is a region near the centre where the temperature is almost constant, and that radiation losses dominate. The equation is of the form $\partial\theta/\partial t = p - q\theta^4$. Thus θ will rise almost linearly at first, and will be asymptotic to the value $\theta = (p/q)^{1/4}$ as $t \to \infty$. If this is only a little greater than θ_c, then because of the asymptotic form of the curve, the time for the fuse to blow will be large and will depend critically on the constants.

51

Let x denote the displacement of the bob, and y that of the hand. Then $m\ddot{x} = -mg(x - y)/l - f\dot{x}$, where l is the length of thread and f is a coefficient depending on drag forces. This gives

$$\ddot{x} + \beta\dot{x} + k^2x = k^2y \tag{1}$$

where $k^2 = g/l$ and $\beta = f/m$. If the movement of the hand is to compensate for the drag forces, y must be in quadrature with x as given by $k^2y = \beta\dot{x}$, and the motion will then be $\ddot{x} + k^2x = 0$. The solution for x is $x = x_0 \exp(ikt)$, and thus we have $k^2y = \beta x_0\, ik \exp(ikt)$, i.e. the amplitude is $y_0 = \beta x_0/k$, or $\beta = ky_0/x_0$.

If the y motion now ceases, (1) becomes $\ddot{x} + \beta\dot{x} + k^2x = 0$, the solution of which is $x = C \exp(-\beta t/2) \exp(i\omega t)$ where $\omega \simeq k$. The amplitude will decay to 1/e of its initial value in a time $\tau = 2/\beta = 2x_0/ky_0$. Inserting numerical values, we obtain $\tau = 255$ s.

A knowledge of the diameters of the bob and the thread would enable one to decide which was contributing most to the drag (probably the bob) and whether the flow was 'streamline' or 'turbulent' —i.e. the value of the Reynolds number. If the former, it would be possible to make an estimate of the viscosity—not very reliable, since Stokes' law applies to a steady state, and here the velocity is changing continually.

52

There are two parts to this problem: (a) to develop the maximum angular velocity about a vertical axis and (b) to remain airborne for as long as possible. The second is simpler: the time available is $t = 2(2h/g)^{1/2}$, where h is the height to which the performer can jump in the circumstances. The first involves the conversion of translational kinetic energy into rotational kinetic energy; since the change is impulsive, the two cannot be assumed to be equal. Considering only motion in a horizontal plane, we have a body of mass m, with a moment of inertia about a vertical axis through the centre of mass of mk^2, and moving with a velocity v. At $t = 0$, one point of this body, a distance r from the trajectory of the centre of mass, is brought to rest: the skater digs her toe into the ice. The result is a rotation about this fixed point with an angular velocity ω given, from the conservation of angular momentum, by $mvr = I\omega_0 = m\omega_0(k^2 + r^2)$, i.e. $\omega_0 = vr/(k^2 + r^2)$. ω_0 has a maximum value of $v/2k$, if r is chosen to be equal to k. If the point of contact of the toe with the ice does not remain stationary, but moves forward with a velocity u, then it is not difficult to show that ω_0 becomes $(v_0 - u)r/(k^2 + r^2)$. Thus the value $vr/(k^2 + r^2)$ is an upper limit for ω_0.

However, by muscular effort, the centre of mass can be moved to lie on the axis of rotation, which is assumed to be fixed in space. This requires a force along the line from the centre of mass to the axis. The angular momentum is thus constant, and the angular velocity increases to $\omega = vr/k^2$. (The energy increases from $m(k^2 + r^2)\omega_0^2/2$ to $mk^2\omega^2/2$, the difference being equal to the work done by the muscles.) There is thus no theoretical limit to the value of ω that can be achieved by increasing r, but anatomical difficulties would limit r to a value roughly equal to k. This would give $\omega = v/k$, which is twice the maximum value obtained above for ω_0.

Once the skater is airborne, the angular momentum must remain constant, but it is still possible to increase the angular velocity if the moment of inertia can be reduced—e.g. by drawing in towards the axis of rotation any arms or legs which have previously been extended. We may thus write $\omega = v/\alpha k$ where, in all likelihood, $1 < \alpha < 2$. The number of turns possible, n, is thus $\omega t/2\pi = v(2h/g)^{1/2}\pi\alpha k$. To obtain a numerical estimate, we set $h = 1$ m, $v = 8$ m s^{-1} and $k = 0.3$ m. This gives a value of n between 2 and 4, depending on the value taken for α—which is not unreasonable. This theoretical model may bear little relation to reality, but it does indicate that (i) k should be kept small, i.e. arms close to the sides or vertically upwards, (ii) it would be more profitable to increase v rather than h, since the latter occurs as a square root, (iii) if it is indeed possible to reduce α, this would pay off very well.

The model completely ignores the fact that the external forces (on the feet) are not acting in the same plane as the motion of the centre of mass.

53

(i) A table of first differences shows that the third entry is anomalous and suggests that 4468 is probably a transcription error for 4408. I would take the risk of making this 'correction'.

(ii) The apparent resistance (V/i) increases with current. Self-heating would give a small temperature rise, proportional to i^2. Try plotting V/i against i^2; a good straight line results, confirming the hypothesis, and gives an intercept of $R_0 = 1.1000 \pm 0.003\ \Omega$ at $i = 0$.

(iii) This should be used as the value of R_0.

(iv) At a temperature T_0, at which $R = R_0$, it would be reasonable to choose a measuring current such that the extra voltage due to heating would be equal to the uncertainties in V, i.e. 1 μV. The graph shows that this current is about 2.2 mA. It would then be necessary to subtract 1 μV from all measured values of V.

At other temperatures, there is no unique 'best' procedure. One possibility would be to choose i so that the temperature rise due to self-heating would be the same as at T_0. This would require that $i_T^2 R_T = i_0^2 R_0$, thus giving i_T if $i_0 = 2.2$ mA. It would then be necessary to increase the 1 μV correction by a factor i_T/i_0 to allow for the change in measuring current, and by a factor $(dR/dT)_T/(dR/dT)_{T_0}$ to allow for the change in resistance. Thus the correction to be subtracted from V would be $(R_0/R_T)^{1/2}(dR/dT)_T/(dR/dT)_{T_0}$ μV. This argument assumes that the temperature rise due to self-heating depends only on i^2R, whereas it will also be affected by the thermal resistance to the surroundings. This will not necessarily be a constant, and in some circumstances may change quite rapidly with temperature.

Instead of choosing i so as to keep the temperature rise due to self-heating constant at all temperatures, an alternative procedure would be to choose it so as to keep the resulting voltage correction constant.

54

(*a*) The annihilation of a particle of rest mass m_1 will make available an energy $\Delta E = m_1c^2$. If this energy is given to another particle of rest mass m_2, which thereby acquires a momentum p, we have the relation

$$(\Delta E + m_2c^2)^2 - p^2c^2 = (m_2c^2)^2.$$

In the particular case envisaged in the problem, $m_1 = m_2 = m$. This leads to the result $p = \sqrt{3}\,mc$. If there are initially N protons and N antiprotons available, this process can be performed N times. If all the N particles ejected can be persuaded to leave the spaceship in the same direction, the effect will be to give a total momentum $\sqrt{3}\,Nmc$ to the ship.

(*b*) If we consider the mutual annihilation of two particles, each of rest mass m, a similar argument shows that the (scalar) total of the momentum thereby produced is $2mc$. There will, in fact, be two photons proceeding in opposite directions (since the vector total must be zero). With the aid of a deep paraboloidal mirror made of a material that can reflect high-energy gamma rays (!) all such photons can be collimated into a beam carrying a momentum $2Nmc$. The spaceship will have an equal momentum in the opposite direction. This is larger than the result of (*a*).

The above arguments are strictly valid only if the increment in the velocity of the spaceship is much less than c. If this is not the case, the total number N can be conceptually subdivided into small groups for which it is true, and the total effect of the momentum increments between successive inertial frames can then be calculated as an integral. The advantage of (*b*) over (*a*) applies at every stage and therefore also overall.

55

Let $\dot{m}$ be the rate of mass transfer ($= 1\ \mathrm{kg\,h^{-1}}$) and h the height through which the sand falls ($= 10$ cm, say). The time of fall is $(2h/g)^{1/2} = t_f$, and the mass in flight at any time in the steady state is $\dot{m}(2h/g)^{1/2}$. This quantity will not be weighed, so there will be a corresponding reduction in the apparent weight of $\dot{m}(2gh)^{1/2}$. However, the velocity of impact is $(2gh)^{1/2}$, and thus the rate of arrival of momentum at the scale pan is $\dot{m}(2gh)^{1/2}$. This will give rise to an increase in the apparent weight which exactly cancels the former reduction. This result could have been established on general grounds, since the centre of mass of the system is not subject to any acceleration in the steady state.

However, we must also consider the possibility of starting and stopping transients. When starting, the mass in flight increases linearly for a time $t_f = 0.14$ s, by which time the apparent weight will have fallen by $\dot{m}(2gh)^{1/2} = 39$ mg. Although the balance could easily register a constant change of this magnitude, its short duration means that more detailed calculations are needed. The moving parts of the balance constitute a system that can oscillate about a stable equilibrium. Its motion will be described by $I\ddot{\theta} + \beta\dot{\theta} + k\theta = 0$, with an obvious notation. Since it is usual for the motion to be critically damped,

$$\beta^2 = 4kI \tag{1}$$

and the appropriate solution of the equation is

$$\theta = (A + Bt)\exp(-\beta/2I)t. \tag{2}$$

The initial condition $\theta = 0$ at $t = 0$ means that $A = 0$. Differentiation of (2) gives

$$\dot{\theta} = B(1 - \beta t/2I)\exp(-\beta/2I)t \tag{3}$$

and so $\dot{\theta}$ at $t = 0$ is equal to B. The transient motion of interest is caused by an out-of-balance force which rises from zero at $t = 0$ to $\dot{m}(2gh)^{1/2}$ at $t = t_f = (2h/g)^{1/2}$, and is zero thereafter. This constitutes an impulse $\dot{m}h$; hence an impulsive couple $\dot{m}hl$ acts on the system, where l is the length of the balance arm. The effect is to produce an initial angular velocity $\dot{m}hl/I$ before any significant movement takes place—i.e. B in equation (3) is equal to $\dot{m}hl/I$. The maximum excursion, when the system comes momentarily to rest, is given by setting $\dot{\theta} = 0$ in (3). It occurs at $t_0 = 2I/\beta$ and its magnitude, from (2), is $2\dot{m}hl/e\beta = \theta_m$.

To obtain numerical values for these quantities, we set $h = l = 10$ cm as before. We take I to be $2Ml^2$, where M is the mass in each pan, i.e. 1 kg. We could find β from (1), if we knew k. This is related to the sensitivity: if an additional load δm on one pan produces a deflection $\delta\theta$, then clearly $k\delta\theta = \delta mgl$. It is stated that the balance can detect $\delta m = 10^{-9}$ kg. For a traditional chemical balance, with a pointer moving over a scale, the limit for $\delta\theta$ would be about 10^{-3} rad; this gives $k = 10^{-6}$. From (1), $\beta = 2.8 \times 10^{-4}$;

this gives $t_0 \simeq 140$ s and the deflection $\theta_m = 8.4 \times 10^{-3}$ rad. This is comfortably larger than the 10^{-3} assumed as the limit of detectability, and so the transient would be detectable. However, a traditional balance capable of taking 1 kg and detecting 10^{-9} kg is rather rare. Modern balances with optical or electronic systems could probably do better. If, for a given δm, we can reduce $\delta\theta$ by a factor f, then k increases by a factor f, and so θ_m decreases by $\sqrt{f}$. Thus the ratio of the actual deflection to the smallest detectable deflection increases by $\sqrt{f}$. The period is also reduced by $\sqrt{f}$.

56

The curve adopted by the wire is a hyperbolic cosine, but since the sag will be very small it will be sufficiently accurate to approximate this by a parabola. We take its equation as $y = \alpha x^2$. Then the slope at the ends is $2\alpha x_0$. If M is the total mass and T is the tensile force, then clearly $Mg = 2T 2\alpha x_0$. But $M = \rho A 2x_0$ (A is the cross-sectional area) and $T = \sigma A$, so $\alpha = \rho g/2\sigma$ and hence $y_0 = \rho g x_0^2/2\sigma$, where y_0 is the height of the ends above the middle. For the values given, $y_0 = 9.87$ cm.

For the second part, the strain $\epsilon(= \delta l/l)$ can be written either as $\beta\delta\theta$ (β is the expansion coefficient) or as σ/E (E is Young's modulus). Taking $\delta\theta = 30°$C, the equality of the two expressions gives $\sigma = 6 \times 10^7$ N m^{-2}, which is only 3% of the breaking stress—thus giving a comfortable margin of safety.

57

If the rubber is incompressible, the volume of material must be a constant, i.e. $r^2 t = r_0^2 t_0$. The excess pressure, when the balloon is inflated, will be $p = 2\sigma t/r$. Substitution of these two results into the equation given for σ leads to the relation

$$p = \frac{2E}{3}\frac{t_0}{r_0}\left[\frac{r_0}{r} - \left(\frac{r_0}{r}\right)^7\right].$$

This has a maximum at $r/r_0 = 7^{1/6} = 1.38$ and goes to zero like r_0/r as r/r_0 goes to infinity. The maximum is rather flat: $1/x - 1/x^7$ has a maximum of 0.62 at $x = 1.38$, and has only fallen to 0.49 at $x = 2$. Thus the increase in σ for higher values of λ causes the curve of $p = p(r/r_0)$ to rise again.

The slope of the curve, i.e. $\partial p/\partial r$, decreases at first, so that it becomes progressively easier to increase the size of the balloon.

This is particularly obvious if the supply of air is from a reservoir whose volume is considerably larger than that of the balloon (e.g. the lungs in the case of a toy balloon). At the maximum a large, discontinuous increase in volume can take place, provided that the reservoir is large enough to make the accompanying decrease in pressure quite small. This phenomenon is well known to parents of small children at Christmas parties.

58

If there were no effects due to the walls of the cube, each neutron would be captured and would give rise to k new neutrons after a time $N\lambda/v$. Thus n neutrons become kn neutrons in a time $N\lambda/v$, i.e. $\mathrm{d}n = (k-1)n$ when $\mathrm{d}t = N\lambda/v$ or $\mathrm{d}n/\mathrm{d}t = nv(k-1)/N\lambda$.

In fact, neutrons will be lost through the sides of the cube. A standard result of gas kinetic theory is that the flux of particles per unit area per unit time is $\rho v/4$, where ρ is the number density of particles. The neutrons will be lost across a total area of $6L^2$ and the density will be n/L^3, where n is the number of neutrons in the cube. Thus, from this effect, $\mathrm{d}n/\mathrm{d}t = -3nv/2L$. Combining the two results, we get

$$\frac{\mathrm{d}n}{\mathrm{d}t} = nv\left(\frac{k-1}{\lambda N} - \frac{3}{2L}\right).$$

Clearly, a chain reaction will take place if $(k-1)/\lambda N > 3/2L$, i.e. if $L > 3\lambda N/2(k-1)$. Inserting the values given, we obtain $L = 375$ m. This enormous size arises through the use of the value of k appropriate to pure ^{238}U. In a real reactor, the value of k would be nearer to 2.0, which gives $L = 15$ m. The critical size can be reduced further by surrounding the core with a layer of material which effectively reflects the neutrons without appreciable absorption.

59

Assuming spherical symmetry, the value of g at a depth δr below the surface will be $G(M - \delta M)/(r - \delta r)^2$, where $M = \frac{4}{3}\pi r^3 \rho_{\mathrm{m}}$ (ρ_{m} is the mean density of the earth) and $\delta M = 4\pi r^2 \delta r \rho_{\mathrm{s}}$ (ρ_{s} is the density of the surface layer). If this is to be greater than GM/r^2, we find that $\rho_{\mathrm{s}} < \frac{2}{3}\rho_{\mathrm{m}}$.

A more general treatment is as follows. We have

$$g(r)=\frac{G}{r^2}\int_0^r 4\pi z^2\rho(z)\,\mathrm{d}z$$

where $\rho(z)$ is the density at radius z. The variation with depth is

$$\frac{\mathrm{d}g}{\mathrm{d}r}=4\pi G\left(\rho(r)-\frac{2}{r^3}\int_0^r z^2\rho(z)\,\mathrm{d}z\right). \qquad (1)$$

Integrating by parts, we obtain

$$\frac{\mathrm{d}g}{\mathrm{d}r}=4\pi G\left(\rho(r)-\frac{2}{r^3}\frac{r^3\rho(r)}{3}+\frac{2}{r^3}\int_0^r \frac{z^3}{3}\frac{\mathrm{d}\rho}{\mathrm{d}z}\,\mathrm{d}z\right).$$

If this is to be negative at $r=R$, we must have

$$\int_0^R z^3\frac{\mathrm{d}\rho}{\mathrm{d}z}\,\mathrm{d}z<-\frac{R^3\rho(R)}{2}$$

where R is the radius of the earth. This clearly requires that $\mathrm{d}\rho/\mathrm{d}z$ be negative and sufficiently large in magnitude over some part, at least, of the range of z.

If we require g to increase with depth at *all* depths (instead of just near the surface, as above), we can find the limiting condition by setting $\mathrm{d}g/\mathrm{d}r\leqslant 0$ in (1). This gives

$$r^3\rho(r)-2\int_0^r z^2\rho(z)\,\mathrm{d}z\leqslant 0.$$

Upon differentiation, we obtain $\rho(r)=k/r$ (k is a constant) as the limit. Any faster increase in ρ with depth will give the desired result.

60

Let us make an obvious change of coordinates: assume that I am at rest, and that the falling rain makes an angle θ with the vertical where $\tan\theta=v_x/v_z$, v_x being my speed and v_z the speed of the rain. We denote the intensity of rainfall, in drops per unit horizontal area per unit time, by n. Consider an element of surface, area S, whose normal lies in the xz plane. Let ϕ be the angle between this surface and the vertical. From the diagram, it is clear that the rain falling on this surface in unit time is

$$nS\sin(\theta+\phi)/\sin(\pi/2-\theta)=nS\sin\phi+nS\cos\phi\tan\theta.$$

$S\sin\phi$ is the vertical projection of S and $S\cos\phi$ is the horizontal projection. The result can therefore be extended to the whole body, provided that no part of my anatomy is sheltered from the rain by some other part. The total rate of wetting, in drops per second, is thus $n(S_z + S_x \tan\theta)$, where S_z is my plan area and S_x is my front elevation. This rate of wetting continues for a time L/v_x, where L is the distance to be travelled. Using the relation $\tan\theta = v_x/v_z$, the total wetting for the trip is $nL(S_z/v_x + S_x/v_z)$. For an order-of-magnitude calculation, take $S_x \simeq 10\ S_z$; v_z is given as $10\ \mathrm{m\,s^{-1}}$ and v_x will not exceed this! Thus the second term dominates, and is independent of my speed, v_x. Clearly, I should move as fast as I can, but it will not have a great deal of effect on my wetness.

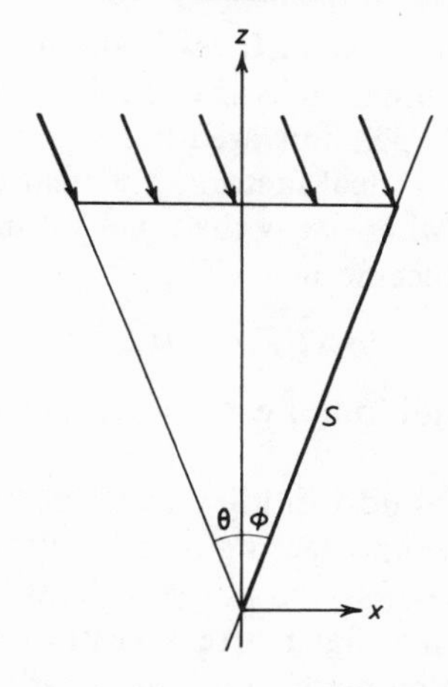

61

The motion consists of two alternating phases: (*a*) rocking on a corner and (*b*) rolling on a side. The transition from rocking to rolling takes place when one corner is vertically above another, i.e. 14 times per revolution. The path of the centre of mass is an arc of a circle, concave downwards, in (*a*) and part of a cycloidal curve, concave upwards, in (*b*). At the transition, there must be no discontinuity, either of displacement or velocity of the centre of mass. On this basis it is possible to carry out a complete analysis of the motion. It is not difficult in principle, but is long and complicated.

Fortunately, for present purposes it is not necessary. The translational velocity will be almost constant throughout the motion: call it v. The only time at which the centre of mass is being accelerated downwards is during the rocking phase. It is then moving on the arc of a circle whose radius is $r = b/2\cos(\pi/14)$ where $b = 3$ cm, as defined in the problem. This expression is approximately equal to $b/2$. The value of the downward acceleration is v^2/r, and this calls for a downward force mv^2/r. Since the contact between coin and table cannot transmit a tensile force, the maximum value of the downward force is mg. Thus the maximum velocity is given by $v^2/r = g$, i.e. $v = 38 \text{ cm s}^{-1}$.

62

If the separation of the two surfaces is small compared with their extent, we may assume that the radiant energy transfer is $\sigma(T_1^4 - T_2^4) \simeq \sigma \Delta T 4T^3$ per unit area per second, and ignore complications due to geometrical factors. The heat transfer via the rain will be $ms(T_1 - T_2)$, where m is the mass falling per unit area per second and s is the specific heat. Thus

$$4\sigma \Delta T T^3 = ms\Delta T$$

for the required condition, i.e. $m = 1.5 \text{ g m}^{-2} \text{ s}^{-1}$—a very light shower!

The effect of thin cloud would be to reduce the transfer of heat by radiation. If it were not acting as a black body, the emissivity of the cloud would be less than unity, at least in some parts of the spectrum. If we assume that the temperature of the cloud base is not significantly affected by the amount of radiation that it receives from the earth, the heat radiated downwards would be less than the σT_1^4 assumed above. There would also be a small contribution of radiation from the upper atmosphere, transmitted through the thin cloud. Since the relevant temperature would probably be less than T_1, then, ignoring complications due to the possible presence of sunshine, this would not suffice to make good the reduction.

63

A proton will go once round the ring in a time $t = L/v$ (L is the circumference, v is the velocity). In this time the total charge will have passed any one point, i.e. the current is $I = Q/t = Qv/L$. Thus $Q = IL/v = 9.5 \times 10^{-5}$ C.

The number of protons is Q/e (e is the electronic charge) and each has an energy eV. The total energy is then QV, with $V = 24 \times 10^9$ V, i.e. the energy is 2.28×10^6 J.

If M is the mass, s is the specific heat and $\Delta\theta$ is the temperature rise, the energy is $Ms\Delta\theta$, giving $\Delta\theta = 60$ K. This assumes that the heat is distributed uniformly throughout the block—which is quite unreasonable since the whole process will last only a few microseconds.

64

For the moon in an orbit of radius r, we have $GMm/r^2 = mr\omega^2$, if $M \gg m$. For the rock of mass μ, in an orbit of radius $r - a$, we have

$$\frac{GM\mu}{(r-a)^2} - \frac{Gm\mu}{a^2} + F = \mu(r-a)\omega^2$$

where F is the force of contact between the rock and the moon. If the rock is to be lifted off, $F = 0$. Eliminating ω between the two equations, we obtain

$$\frac{M}{m} = \frac{r^3}{a^3}\frac{(r-a)^2}{3r^2 - 3ra + a^2} \simeq \frac{1}{3}\left(\frac{r}{a}\right)^3$$

as required.

65

In metallic conduction the electrons are scattered by the ions under the influence of random thermal motion. The averaged effect of these collisions may be represented by a force opposing the motion and proportional to the drift velocity $\bar{v}$. The equation of motion for an electron (mass m, charge e) in an applied electric field E is then

$$m\,\mathrm{d}\bar{v}/\mathrm{d}t + k\bar{v} = eE. \tag{1}$$

This has the steady-state solution $\bar{v} = eE/k$. If the number density of electrons is N, then the current density is

$$j = Ne\bar{v} = Ne^2E/k = \sigma E \tag{2}$$

where $\sigma \equiv Ne^2/k$ is the conductivity.

In the experiment described, consider an element of wire moving with a constant speed u. The drift velocity in a frame of reference moving with the wire will be zero, since no applied field is present. It is therefore u in the laboratory frame. After the wire is suddenly brought to rest, at $t=0$, say, the collisions cause the drift velocity to decay according to $m\mathrm{d}\bar{v}/\mathrm{d}t + k\bar{v} = 0$, i.e., since $\bar{v} = u$ at $t = 0$, $\bar{v} = u\exp(-kt/m)$. The corresponding current density is $j = Ne\bar{v}$, and the total current that flows is

$$Q = A\int_0^\infty j(t)\,\mathrm{d}t = AmNeu/k$$

where A is the cross-sectional area. We now substitute for k from (2) and introduce the resistance of the wire, $R = l/\sigma A$ (l is the length), and find $Q = ulm/Re$.

With the values given, $e/m = 2\times 10^{11}\ \mathrm{C\,kg^{-1}}$.

66

Consider the motion of a thin layer of water lying at a distance between y and $y+\delta y$ from the solid surface. It can be considered to be moving with constant velocity, so the combined effects of gravity and viscosity must give zero resultant force. This gives $\eta\partial^2 v/\partial y^2 + \rho g\sin\alpha = 0$, where v is the velocity at y. This can be integrated immediately, with $v=0$ at $y=0$ and $\partial v/\partial y = 0$ at $y=d$, to give

$$v = \frac{\rho g \sin\alpha}{\eta}\left(yd - \frac{y^2}{2}\right).$$

The volume rate of flow is then $\int_0^d v\,\mathrm{d}y$, which gives the desired result, namely $q = (\rho g\sin\alpha)d^3/3\eta$. By considering the flow into, and out of, a slab of thickness δx perpendicular to the flow direction, we obtain

$$\frac{\partial d}{\partial t}\,\delta x = -\frac{\partial q}{\partial x}\,\delta x$$

which leads immediately to the equation quoted. The correctness of the solution given can be shown by differentiating it with respect to d, first with t constant, and then with x constant.

We obtain

$$\eta\,\frac{\partial x}{\partial d} - 2\rho g t d\sin\alpha = F'(d)$$

and

$$0 - \rho g\sin\alpha\left(2dt + d^2\,\frac{\partial t}{\partial d}\right) = F'(d).$$

Hence

$$\eta \frac{\partial x}{\partial d} + \rho g d^2 \sin\alpha \frac{\partial t}{\partial d} = 0$$

from which the desired result follows if we multiply by $(\partial d/\partial x)(\partial d/\partial t)$.

If we now choose $F(d) = Kd$, where K is a 'constant' with the dimensions of a viscosity, then, at $t = 0$, $d = \eta x/K$, which meets the boundary conditions specified in the example. The equation for d then becomes $\rho g t d^2 \sin\alpha + Kd = \eta x$. When t is large enough, the first term will dominate the second; for $x = l$, $d = (\eta l/\rho g t \sin\alpha)^{1/2}$. With the numerical values given, $d = 2.06 \times 10^{-6}$ m.

67

For a parallel-plate capacitor, we have $C = \epsilon\epsilon_0 A/d$ where A and d are the area and separation of the plates. When a current is flowing in a circuit consisting only of a capacitance C and an inductance L, then, if Q is the charge on the capacitor at any time, we have

$$\frac{\mathrm{d}^2Q}{\mathrm{d}t^2} + \frac{Q}{LC} = 0. \quad (1)$$

If the plates of the capacitor are vibrated sinusoidally, we can write

$$d = d_0 + d_1 \sin\Omega t$$

so that

$$1/C = (d_0 + d_1 \sin\Omega t)/A\epsilon\epsilon_0 = 1/C_0 + \eta \sin\Omega t$$

say. Equation (1) then becomes

$$\frac{\mathrm{d}^2Q}{\mathrm{d}t^2} + \frac{Q}{LC_0} = -(\eta Q/L)\sin\Omega t. \quad (2)$$

We write $1/LC_0 = \omega^2$, where ω is the resonant frequency with the plates at rest. To solve (2) we first obtain a zero-order approximation by setting $\eta = 0$. The result is $Q = Q_0 \sin\omega t$. Using this solution on the right-hand side of (2), we obtain the first-order approximation

$$\frac{\mathrm{d}^2Q}{\mathrm{d}t^2} + \omega^2 Q = -(\eta Q_0/L)\sin\omega t \sin\Omega t$$

$$= \frac{\eta Q_0}{2L}\,[\cos(\omega - \Omega)t - \cos(\omega + \Omega)t]. \quad (3)$$

In general, this gives sinusoidal solutions for Q with frequencies $\omega - \Omega$ and $\omega + \Omega$. However, if $\Omega = 2\omega$, the first term in (3) becomes $\cos(-\omega t) \equiv \cos \omega t$ and we have the special case in which the solution involves $t \sin \omega t$ and is, in fact

$$Q = \frac{\eta Q_0}{4\omega L}\, t \sin \omega t + \frac{\eta Q_0}{16\omega^2 L} \cos 3\omega t.$$

The current $I = \mathrm{d}Q/\mathrm{d}t$ includes a (small) term in $3\omega t$ and also a component $(\eta Q_0/4\omega L)(1 + \omega^2 t^2)^{1/2} \sin(\omega t + \phi)$, with $\tan \phi = \omega t$. After a few cycles, the amplitude becomes proportional to t, and the phase has changed from 0 to $\pi/2$.

68

A non-zero longitudinal polarisation of a particle corresponds to a non-zero expectation value $\langle \boldsymbol{S} \cdot \boldsymbol{p} \rangle$, where $\boldsymbol{S}$ is the spin vector of the particle and $\boldsymbol{p}$ is its momentum. These are pseudo (or axial) vector and true (or polar) vector quantities respectively, so $\boldsymbol{S} \cdot \boldsymbol{p}$ is a pseudo-scalar. The strong nuclear interaction conserves parity, and parity conservation requires that the expectation of a pseudo-scalar such as $\boldsymbol{S} \cdot \boldsymbol{p}$ must be zero. Hence there can be no longitudinal polarisation.

On the other hand, the polarisation transverse to the scattering plane is measured by $\langle \boldsymbol{S} \cdot \boldsymbol{p}_{\mathrm{a}} \wedge \boldsymbol{p}_{\mathrm{b}} \rangle$, where $\boldsymbol{p}_{\mathrm{a}}$ and $\boldsymbol{p}_{\mathrm{b}}$ are the momenta of projectile a and product b. Since $\boldsymbol{p}_{\mathrm{a}} \wedge \boldsymbol{p}_{\mathrm{b}}$ is a pseudo-vector, $\boldsymbol{S} \cdot \boldsymbol{p}_{\mathrm{a}} \wedge \boldsymbol{p}_{\mathrm{b}}$ is a true scalar and its expectation value does not vanish as a consequence of parity conservation. Hence a transverse polarisation is allowed.

The event of 'left scattering with up polarisation' is related to an event 'right scattering with down polarisation' by rotation through π about the incident momentum $\boldsymbol{p}_{\mathrm{a}}$ as axis. Rotational invariance of the interaction requires that the numbers of these two events be equal.

69

In statistical mechanics, entropy is frequently introduced by considering arrangements of distinguishable elements, like a pack of cards. Since there are 52 cards in a pack (excluding the Joker!) it would seem that there are 52! ways of arranging them, and the definition of entropy would suggest that $S = k_{\mathrm{B}} \ln 52!$ If we now choose a *particular* arrangement, then there is only one way to order the cards: $S = k_B \ln 1 = 0$ and hence the entropy has decreased.

However, entropy is concerned with the states of physical systems—states which are accessible, within certain constraints, as a result of the thermal energy of the system. The entropy is then a function of temperature as well as other variables, i.e. $S = S(T, X)$. To achieve a cooling, the entropy is reduced, at constant temperature, via the variable X (e.g. mechanical work, magnetic field) and then, keeping the entropy constant, X is changed and the temperature is thereby reduced. Since it is difficult to imagine a pack of cards, left to its own devices, shuffling itself through all possible arrangements at *any* temperature, the concept of entropy is inappropriate.

It would seem to be inadvisable to invest in a refrigeration system based on such a principle.

70

Let us choose the x axis along the direction of the field. We then write the field as $(E_0 + E_1 x)\cos\omega t$, and consider the motion of an electron situated initially at $x = 0$. This will be described by

$$m\ddot{x} = e(E_0 + E_1 x)\cos\omega t. \qquad (1)$$

For a zero-order approximation, we neglect $E_1 x$ compared with E_0; hence $m\ddot{x}_0 = eE_0 \cos\omega t$, i.e. $x_0 = -eE_0 \cos\omega t / m\omega^2$. For the next approximation, we write $x = x_0 + x_1$, which defines x_1. Substitution in (1) gives

$$m\ddot{x}_0 + m\ddot{x}_1 = e[E_0 + E_1(x_0 + x_1)]\cos\omega t.$$

But $m\ddot{x}_0 \equiv eE_0 \cos\omega t$, and therefore $m\ddot{x}_1 = eE_1(x_0 + x_1)\cos\omega t$. We now neglect x_1 compared with x_0 in this expression, giving

$$m\ddot{x}_1 = eE_1 x_0 \cos\omega t = -(e^2 E_0 E_1 / m\omega^2)\cos^2\omega t. \qquad (2)$$

This could be solved, and gives for x a result of the form $x_1 = pt^2 + q\cos 2\omega t$. Thus in addition to oscillatory motions, there is a steady force, producing the accelerated motion pt^2. To find p, we merely take the time average of (2) over one cycle to obtain the force $-e^2 E_0 E_1 / 2m\omega^2$. This could be written as $(-e^2/4m\omega^2)\partial(E^2)/\partial x$—neglecting a small term in E_1^2—where the negative sign shows that the force is always directed towards a place where the field is weaker, and eventually towards a minimum where $\partial(E^2)/\partial x = 0$.

71

Consider first an equilibrium situation, where there is no ventilation and a definite amount of liquid is present. The amount of mercury needed to reach the hazardous threshold quoted in a room of 300 m^3 capacity is 15 mg. For benzene, the amount is about 9 g. Thus 'a few grams' of mercury would easily be sufficient to produce dangerous concentrations—in time—but the verdict on benzene is more doubtful.

However, a more realistic treatment takes account of the rate of ventilation, the rate of evaporation and, perhaps, the rate at which the liquid is replenished. The rate of evaporation can be estimated in two ways.

(i) Consider a liquid in contact with its saturated vapour. In dynamic equilibrium, the rate of evaporation will be equal to the rate of impact of vapour molecules on the liquid surface. From kinetic theory this is $n\bar{v}/4$ atoms per unit area per second, where n is the number of atoms per unit volume and $\bar{v}$ is given by $(3k_BT/m)^{1/2}$, m being the mass of an atom. The mass rate of evaporation is thus $\frac{1}{4}mn(3k_BT/m)^{1/2}$. In the gas law, $pV = RT$, V is the volume occupied by Z atoms at a pressure p (Z is Avogadro's number). Thus $n = Z/V$ and the evaporation rate is

$$\frac{mZ}{4V}\left(\frac{3k_BT}{m}\right)^{1/2} = \frac{mZ}{4}\frac{p}{RT}\left(\frac{3k_BT}{m}\right)^{1/2} = \frac{p}{4}\left(\frac{3m}{k_BT}\right)^{1/2}.$$

Setting p equal to the saturation vapour pressure for mercury, we obtain a value of approximately 8×10^{-5} kg m^{-2} s^{-1}. This is likely to be an overestimate.

(ii) An alternative, empirical estimate can be made from the observation that a small droplet of mercury, lying on a bench, lasts for a long time. If we assume that the rate of evaporation per unit area is constant, even for very small drops, then it is easy to show that the radius decreases linearly with time, and that if a drop with an initial radius r_0 finally disappears in a time τ, then the mass rate of evaporation per unit area is $\rho r_0/\tau$ (ρ is the density). If we assume that a millimetre droplet of mercury has not completely disappeared in two weeks, then we can calculate an upper limit to the rate of 6×10^{-6} kg m^{-2} s^{-1}. This is a good deal smaller than the kinetic theory estimate; the discrepancy is discussed further below. For order-of-magnitude calculations we may take 10^{-5} kg m^{-2} s^{-1}, i.e.

$10^{-3}\ \mathrm{mg\,cm^{-2}\,s^{-1}}$. 'A few grams' of mercury will have a surface area of the order of $1\ \mathrm{cm^2}$, so that the rate will be about $10^{-3}\ \mathrm{mg\,s^{-1}}$.

This must be set against the rate of ventilation. Standard practice calls for three air changes per hour, and a room which is 'not specially ventilated' is unlikely to have more. To present a worst case, assume one change per hour. In the present example, this means $0.88\ \mathrm{m^3\,s^{-1}}$. If the mixing is complete, this would give an average concentration of mercury of $0.012\ \mathrm{mg\,m^{-3}}$. This is indeed less than the danger level of $0.05\ \mathrm{mg\,m^{-3}}$, but the factor of safety ($\simeq 4$) is not large enough to allow comfortably for possible errors in the calculated rate.

Calculations on benzene can be carried out along similar lines. The kinetic theory estimate of the rate of evaporation is $22\ \mathrm{kg\,m^{-2}\,s^{-1}}$, i.e. $2.2\ \mathrm{g\,cm^{-2}\,s^{-1}}$—which is very large, and quite unrealistic. An empirical estimate, as in the case of mercury, suggests a value smaller by a factor of 10^3. Two effects will contribute to the discrepancy: the surface layers will be cooled by the evaporation and—more importantly—the surface will be covered by a layer of saturated vapour. Diffusion of vapour through this, into free air, is likely to be the rate-determining process. The theoretical rate might be approached in the case of evaporation into a high vacuum, with sufficient heat supply to provide the latent heat of evaporation (almost 1 kW per $\mathrm{cm^2}$ of surface!).

In the circumstances of the problem, the evaporation rate will clearly be very high, and a different approach is more helpful. We saw above that about 9 g of benzene are sufficient to produce a dangerous concentration, and the discussion suggests that, if this were in an open dish, it could all evaporate in an hour. If this were an isolated event, the ventilation would keep the concentration below the danger level over most of the volume for most of the time. However, if the nature of the work in progress required that the benzene be replenished frequently, then there would be cause for alarm.

As a consultant physicist I would—true to type—ask for more information before giving an opinion. I would need data on the actual rate of ventilation and on the nature of the operations involving benzene; and I would perform some simple measurements on the rate of evaporation of benzene. I would be fairly confident that the danger from mercury was not serious—but it would be very easy to observe the rate at which droplets disappeared, and this could be done to confirm the opinion.

72

The thermal conductivity of a gas is $k \simeq n k_B \lambda \bar{c}/2$ (n is the number of molecules per unit volume and λ is the mean free path). We also have $\lambda = 1/\pi n a^2$ (a is the molecular diameter) and $\bar{c} = (3k_B T/m)^{1/2}$. These relations give $k = cT^{1/2}$, where c is a constant which can be evaluated if we assume a value for a: 2×10^{-10} m would be reasonable. The rate of heat flow is $\dot{Q} = kA\, \mathrm{d}T/\mathrm{d}x$, where A is the cross-sectional area of the column. In the steady state, $\dot{Q}$ is independent of x; hence $\dot{Q}l = \frac{2}{3}cA(T_1^{3/2} - T_0^{3/2})$, where l is the length of the column and T_1, T_0 are the temperatures at the top and bottom respectively. Inserting the values, we obtain $\dot{Q} = 2.5 \times 10^{-3}$ J s^{-1}; the boil-off rate is about 3 cm^3 per hour.

The above answers the question asked, but it is perhaps worthwhile to point out that other effects—such as heat conduction down the walls of the Dewar—may be equally important.

73

The basic assumption is that the distribution of disintegration energies is fairly flat and—in the limit—that the number per unit energy range is a constant. Then, since $E \propto \tau^{-1/5}$, $\mathrm{d}E \propto \tau^{-1.2}\,\mathrm{d}\tau$, and hence the distribution in terms of τ is given by $\mathrm{d}N = c_1 \tau^{-1.2}\,\mathrm{d}\tau$, where c_1 is a constant.

Consider now only one particular decay process with a mean lifetime of τ_n. The probability of observing such a decay in a time $\mathrm{d}t$ is

$$P(t)\,\mathrm{d}t = (1/\tau_n)\exp(-t/\tau_n)\,\mathrm{d}t.$$

For all the possible decay processes, the probability is found by summing over all values of n, with each term weighted according to the number of the corresponding species present. This weighting factor is just the expression found above,

$$\mathrm{d}N = c_1 \tau^{-m}\,\mathrm{d}\tau \qquad (1)$$

with $m = 1.2$. For a continuous range of τ, this sum can be expressed as an integral, i.e.

$$P(t)\,\mathrm{d}t = \mathrm{d}t \int_0^\infty c_1\tau^{-m}\left(\frac{1}{\tau}\right)\exp\left(\frac{-t}{\tau}\right)\,\mathrm{d}\tau$$

$$= \mathrm{d}t\,c_1 \int_0^\infty \tau^{-(1+m)} \exp\left(\frac{-t}{\tau}\right)\,\mathrm{d}\tau.$$

We now introduce a new variable x, defined by $x = \tau/t$. Then

$$P(t)\,\mathrm{d}t = \mathrm{d}t\,c_1 t^{-m} \int_0^\infty x^{-(1+m)} \exp\left(\frac{-1}{x}\right)\,\mathrm{d}x.$$

For a given m, the integral is just a number, so that $P(t)\,\mathrm{d}t = c_2 t^{-m}\,\mathrm{d}t$ (c_2 is a constant), which is equal to $c_2 t^{-1.2}\,\mathrm{d}t$ in this case, as required.

(By way of a bonus we may note that if $m > 1$, expression (1) becomes infinite if integrated over all values of τ from 0 to ∞. Thus it cannot be completely correct, but must be modified at small values of τ. It can be shown that this change does not alter the main character of the results.)

74

Let V denote the voltage of the source and R the resistance of the motor. Then the current when the motor is stationary is V/R. Let B be the back EMF when running at full speed. The current will then be $(V - B)/R$, and this is stated to be $V/10R$. This gives $B = 0.9V$. The power dissipated is $I^2R = (V/10R)^2R = 0.01\,V^2R$. When running at half speed, the back EMF will be halved, since the same coil is rotating at half the speed in the same magnetic field. It will thus be $0.45\,V$. Since we are told that the speed is proportional to the torque, and since the torque is proportional to the current, in the same field, the current will also be halved when running at half speed. It will thus be $V/20R$.

(*a*) If the speed reduction is caused by a series resistor r, we have

$$\frac{V}{20R} = \frac{V - 0.45\,V}{R + r}$$

from which $r = 10R$, and the power dissipated is $I^2R = (V/20R)^2(R + r) = 0.0275\,V^2/R$. However, most of this will be dissipated in the series resistor and only $(V/20R)^2R = 0.0025\,V^2/R$ will appear in the motor.

(b) If the transistor is used, the mean current again has the value appropriate to half-speed running, i.e. $V/20R$, but this is achieved by allowing the full current $(V-0.45V)/R$ to flow for only a fraction f of the time. Thus

$$\frac{V}{20R}=\frac{V-0.45V}{R}f$$

whence $f=\frac{1}{11}$. The power dissipated is now $[(V-0.45V)/R]^2R$ for a fraction f of the time, and zero for the remainder. The mean power is thus $(0.55V^2)/R\times\frac{1}{11}=0.0275V^2/R$. This is the same as the total dissipation in part (a), but now it all appears in the motor. Compared with the full-speed value, method (a) reduces the dissipation in the motor to $\frac{1}{4}$ while method (b) increases it by a factor of 2.75.

75

We consider the field due to Q when at a distance d above the surface, and then let $d\to 0$. The method of electrical images is standard textbook work. For a charge near to a plane dielectric boundary, there are two images, one (Q_1) at a distance d below the surface and the other (Q_2) coinciding with Q. The field at any point outside the oil is the same as that which would be produced (with no oil present) by Q and Q_1. The field at any point inside the oil is the same as would be produced by Q_2 if the oil extended everywhere to infinity. The values of Q_1 and Q_2 are found by writing down the boundary conditions at the interface. The results are

$$Q_1=-\frac{\epsilon-1}{\epsilon+1}Q \qquad Q_2=\frac{2\epsilon}{\epsilon+1}Q.$$

As $d\to 0$, the field outside tends to the field, *in vacuo*, of a charge $Q+Q_1$, i.e.

$$E=\frac{1}{4\pi\epsilon_0}\frac{2Q}{\epsilon+1}\frac{1}{r^2}$$

at a distance r. The field inside the oil is that of Q_2, i.e.

$$E=\frac{1}{4\pi\epsilon_0\epsilon}\frac{2\epsilon Q}{\epsilon+1}\frac{1}{r^2}$$

which is the same.

It is obvious that either of the two possible symmetrical positions are positions of equilibrium. From a consideration of the effects of a small displacement from equilibrium, it is clear that positions with the charges q at 45° to the horizontal and vertical are unstable, and it follows that the others, with the q in the horizontal and vertical positions, must be stable. At intermediate positions there will be a net couple—but no perpetual motion!

76

There are four essentially different possible equilibrium configurations, as shown in the diagram. Consider (*a*) first. Insert an imaginary, massless, gas-tight, frictionless piston to divide the cylinder into two parts, A and B, and consider an infinitesimal movement of the piston to the right. This will cause the bubble on B to grow, and so decrease its radius. Thus the pressure in B $(=4\gamma/R)$ will be larger than its original value. Similarly the pressure in A will have been reduced, so the pressure difference acts to restore the piston to its original position, i.e. the situation is stable. Similar arguments applied to (*b*) and (*c*) show that the former is stable and the latter unstable. Case (*d*) is more subtle. The curvature of both films decreases when the piston moves from A to B, i.e. the pressure on both sides is reduced, and it is not clear which is reduced by the greater amount.

If we note that, in this case, the two spherical caps together form a complete sphere, the algebra of a formal approach becomes tractable. The pressure is $p = \Pi + 4\gamma/R$ (Π is the atmospheric pressure) and the total volume of gas is $V = V_0 + 4\pi R^3/3$ (V_0 is the volume of the cylinder). The total energy is thus $E = pV + 8\pi R^2\gamma$. The equilibrium configuration can be found by setting $\partial E/\partial R = 0$ and solving for R. The equation is a quartic, but fortunately there is no need to solve it. The solution will correspond to a stable

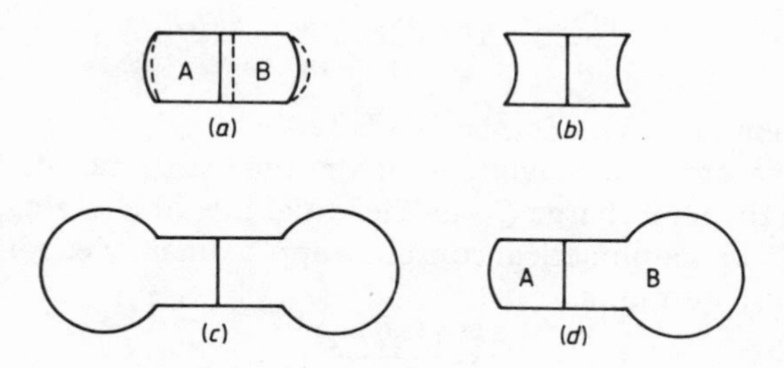

equilibrium if E is a minimum, i.e. if $\partial^2 E/\partial R^2$ is positive. The expression for $\partial^2 E/\partial R^2$ is of the form $aR + b + c/R^3$, where all the coefficients are combinations of physical magnitudes already defined, and must be positive. Thus $\partial^2 E/\partial R^2$ is positive for all positive values of R, and the equilibrium, wherever it may be, is stable.

77

(a) The number of electrons hitting unit area of the dust particle in unit time is $\frac{1}{4}n\bar{v}_e$, where $\bar{v}_e$ is the mean velocity of the electrons. This is a standard result in kinetic theory. A similar expression appears in the case of protons, with mean velocity $\bar{v}_p$. Hence the initial rate of accumulation of charge on the particle will be $\pi r^2 ne(\bar{v}_e - \bar{v}_p)$, where r is its radius. If the temperature T is sufficiently high, we can use the equipartition theorem to give

$$\tfrac{1}{2}m_e\overline{v_e^2} = \tfrac{1}{2}m_p\overline{v_p^2} = \tfrac{3}{2}k_B T.$$

Since $\bar{v}_e \simeq (\overline{v_e^2})^{1/2}$, the rate of accumulation of charge is

$$\frac{dQ}{dt} \simeq \pi r^2 ne(3k_B T)^{1/2}(m_e^{-1/2} - m_p^{-1/2}).$$

Clearly the particle will become negatively charged.

(b) When the particle is charged, approaching protons will be accelerated and electrons will be retarded. When a balance is eventually reached, it can be described by equating the impact velocities of the protons and the electrons. Since the potential at the surface of a sphere of radius r, carrying a charge Q, is Q/r, the impact velocity of an electron will be $[2(m_e v_e^2/2 - eQ_m/r)/m_e]^{1/2}$, with a similar expression for a proton, but with a positive sign. Q_m is the total charge at this stage. Equating the two expressions, and setting the mean kinetic energies remote from the dust particle equal to $\frac{3}{2}k_B T$ as before, we obtain

$$\frac{eQ_m}{r} = \frac{3k_B T}{2}\frac{m_p - m_e}{m_p + m_e} \simeq \frac{3k_B T}{2}.$$

The potential is thus $Q_m/r = 3k_B T/2e$.

(c) The order of magnitude of the time scale can be found by dividing the final charge Q_m by the initial rate of charging, as given above. If we set numerical constants approximately equal to 1, and $m_p \gg m_e$, we obtain

$$\tau \simeq \frac{(k_B T m_e)^{1/2}}{nre^2}.$$

78

The simple lens formula gives $v = uf/(u - f) \simeq f + f^2/u + \ldots$. If the radius of the objective lens or mirror is R, then if the observing screen is misplaced by δv, the radius of the 'geometrical' image of a point source is $R\delta v/v$. In the present case, $v \simeq f$ since the star is effectively at infinity, and $\delta v = f^2/u$: thus the image size is $Rf/u = Rf/3.8 \times 10^8$. Only if this quantity is larger than the size of the image arising from other effects would it be necessary to refocus. These other effects are as follows.

(i) *Atmospheric turbulence.* This limits the resolution to about 1 arc second, i.e. to about 5×10^{-6} rad. The corresponding image size is $5 \times 10^{-6} f$, and the out-of-focus image, Rf/u, is bigger than this if $R > 2 \times 10^3$ m.

(ii) *Grain size of the emulsion.* This may be taken to be $10\ \mu m = 10^{-5}$ m. The limit is thus given by $Rf/3.8 \times 10^8 = 10^{-5}$. If the lens is operating at $f/5$ (which is reasonable) this gives $R > 27$ m as the criterion for refocusing.

(c) *Diffraction.* The radius of the Airy disc is $1.22\lambda f/R$; for lenses operating at $f/5$, this becomes 6.1λ, independent of the absolute size. Taking $\lambda = 5 \times 10^{-7}$ m, we find that the out-of-focus image size is larger than the Airy disc if $R > 15$ m.

Thus the resolution is normally limited by atmospheric turbulence, but even if the effects of turbulence and grain size could be eliminated, we may conclude that for any real telescope, with $R < 15$ m, refocusing would not be necessary, because diffraction effects are more serious than the out-of-focus blurring.

79

For convenience, we take the temperature of the surroundings as the zero of the temperature scale, and assume that the milk is at this temperature. All specific heats are taken as unity; this will introduce no great error. We also assume that the rate of loss of heat to the surroundings is proportional to the temperature excess above the surroundings, i.e. $\delta Q = m\delta\theta = -\alpha\theta\delta t$, where m is a mass and α is a constant depending upon the conditions. This means that cooling while waiting is exponential, of the form $\theta = \theta_0 \exp(-\alpha t/m)$. The change of temperature on mixing is found by elementary methods.

We then consider two sequences, starting from the same state.

(a) Add milk and then allow to stand for a time t. The final temperature is readily found to be

$$\theta_a = \frac{M\theta_0}{M+m} \exp\left(\frac{-\alpha_a t}{M+m}\right)$$

where M, m are the masses of coffee and milk respectively.

(*b*) Allow to stand for the same time and then add the milk. We find

$$\theta_b = \frac{M\theta_0}{M+m} \exp(-\alpha_b t/M).$$

The difference $\theta_a - \theta_b$ can be simplified if we are justified in expanding the exponentials: this is very plausible. We find

$$\theta_a - \theta_b = \frac{M\theta_0 t}{M+m}\left(\frac{\alpha_b}{M} - \frac{\alpha_a}{M+m}\right).$$

An obvious assumption that we can use to simplify this expression further is that the rate of loss of heat is independent of the mass of liquid, i.e. $\alpha_a = \alpha_b = \alpha$, say. This gives

$$\theta_a - \theta_b = \frac{m\theta_0 \alpha t}{(M+m)^2}$$

which is positive, i.e. procedure (*b*) gives the lower final temperature. For a little greater refinement, we must consider the mechanism of the loss of heat. Radiation losses will be small at these low temperatures. The conduction to the bench will be small because of poor conductivities and a small area of contact. Convection losses will dominate, and an important feature here will be evaporation from the surface. (The well known trick of putting the saucer on *top* of the cup to keep the tea warm confirms this hypothesis!) Since, with the larger volume of liquid, the surface will be closer to the top of the mug, we might expect α_a to be larger than α_b. As a most improbable—but mathematically convenient—extreme case we could set α proportional to the mass of liquid. This would clearly give $\theta_a - \theta_b = 0$. It seems likely, therefore, that any more reasonable assumption would not alter the conclusion that $\theta_a - \theta_b > 0$.

It t were so large that it was not justifiable to expand the exponential in series, the conclusion would be the same, but less obvious; but—except as an academic exercise—the problem would then have ceased to exist.

80

There are two possible regimes that can exist. The first, and most likely, occurs when the diffusion of vapour along the tube is the rate-determining process. The assumption is that the partial pressure of the vapour at the meniscus is equal to the saturation vapour pressure. We can also, plausibly, assume that the vapour pressure at the open end is zero. For simplicity, we also assume that everything is at room temperature when the experiment begins, with the tube full. If, as suggested, all changes are slow, then the pressure gradient of the vapour at time t will be p_s/x, where p_s is the saturation vapour pressure and x is the distance of the meniscus below the open end. The rate of mass transfer by diffusion will then be equal to cAp_s/x, where A is the cross-sectional area of the bore and c is a constant. This loss must be made up by evaporation, and so must also be $\rho A\,\mathrm{d}x/\mathrm{d}t$ (ρ is the density). Thus $\mathrm{d}x/\mathrm{d}t = (cp_s/\rho)(1/x)$, i.e. $x \propto t^{1/2}$ if $x = 0$ at $t = 0$.

If this state of affairs is to exist, the supply of heat must be sufficient to evaporate liquid at the necessary rate. The liquid at the meniscus will be cooled by the evaporation, so there will be a temperature gradient across the walls of the tube. There will be a critical value of the thermal resistance of this path which just gives the necessary heat flow. If the resistance is less than this critical value, the rate of evaporation will not be affected, since the liquid temperature cannot rise above room temperature, and so the vapour pressure cannot exceed p_s. If the thermal resistance is greater than the critical value, however, liquid will not evaporate fast enough to keep pace with the rate of diffusion, and hence the rate of descent of the meniscus will be reduced. With a given experimental arrangement, this second regime is most likely to arise when the rate of mass transfer by diffusion is high—i.e. in the early stages when the tube is almost full. When t is small, therefore, x will increase more slowly than $\mathrm{t}^{1/2}$. It is at this stage also that the assumption of a quasi-steady state is likely to break down.

81

The word 'rapidly' used in this context suggests a speed of the order of $1\ \mathrm{m\,s^{-1}}$. We can thus estimate the Reynolds number $R = \rho vl/\eta$. The density is $\rho = 10^3\ \mathrm{kg\,m^{-3}}$, the velocity is $v = 1\ \mathrm{m\,s^{-1}}$, the linear dimension is $l = 3 \times 10^{-3}$ m and the viscosity is $\eta = 10^{-3}\ \mathrm{kg\,m^{-1}\,s^{-1}}$.

This gives $R \simeq 3 \times 10^3$. The critical value for the transition to turbulent flow is about 10–100, and the calculated R is comfortably larger than this. We can therefore take the drag force to be $c\rho A v^2$, where c is a constant of order unity depending on the shape of the body concerned and A is the cross-sectional area, of order l^2.

The propelling force arises from the reduction in surface tension at the tail. If it were reduced to zero, the unbalanced forward force would be of order γl, where γ is the surface tension. This is an upper limit: we write it as $f\gamma l$, with $f < 1$. The steady-state velocity is thus given by $f\gamma l = c\rho l^2 v^2$. With $c \simeq f \simeq 1$ this gives $v \simeq 0.2\ \mathrm{m\,s^{-1}}$. This is still big enough to make R larger than the critical value.

If, as implied, the motion is confined to the surface, then the 'wave-making resistance', rather than the drag force as above, might turn out to be the dominant factor.

82

We denote the radius and mass of the earth by R and M, and the radius of the satellite orbit by r. The velocity of the satellite, v, is given by $v^2 = GM/r = gR^2/r$. Time dilation reduces the clock rate by $(1 - v^2/c^2)^{1/2} \simeq 1 - gR^2/2c^2r$. The difference in gravitational potential between the satellite and the surface of the earth is $\Delta\phi = GM(1/R - 1/r) = gR^2(1/R - 1/r)$. This will increase the clock rate by $(1 + \Delta\phi/c^2) = 1 + gR^2(1/R - 1/r)/c^2$. The combined effect is that the satellite clock rate differs from the earth clock rate by a factor $1 + gR^2(1/R - 3/2r)/c^2$. Thus the satellite clock will lose for $r < 3R/2$, but will gain at greater heights. The biggest possible loss rate will be when $r \simeq R$; its value is $gR/2c^2 = 3.8 \times 10^{-10}$, which is about 12 ms per year. For a typical orbit at a height of about 200 km, the rate would be about 11 ms per year.

The best accuracy currently attainable with caesium clocks is of the order of 10^{-12}, while the overall accuracy under laboratory conditions is somewhat less, say 10^{-11}. Thus the effect in an ordinary satellite ($\simeq 3 \times 10^{-10}$) should be readily detectable.

83

The axial component of the magnetic field, $B_\parallel$, will induce a current in the ring, which will interact with the radial component of the field, $B_\perp$—i.e. the 'leakage flux'—to produce an axial force.

The flux through the ring is

$$\int B_{\parallel}\, \mathrm{d}A \sin \omega t = \bar{B}_{\parallel} A \sin \omega t.$$

The integration is over the area of the ring, A, although most of the flux is in the iron. The induced EMF is $\omega \bar{B}_{\parallel} A \cos \omega t$, and the resulting current I is $\omega \bar{B}_{\parallel} A \cos(\omega t - \phi)/(R^2 + \omega^2 L^2)^{1/2}$, where R is the resistance and L is the inductance of the ring, and $\tan \phi = \omega L/R$. The force is $F = IB_{\perp} 2\pi r$, where r is the radius of the ring. $B_{\perp}$ also varies as $B_{\perp} \sin \omega t$. The resulting force is thus

$$\frac{\omega \bar{B}_{\parallel} A B_{\perp} \cos(\omega t - \phi) \sin \omega t\, 2\pi r}{(R^2 + \omega^2 L^2)^{1/2}}.$$

The trigonometric terms can be written as $\sin(2\omega t - \phi) + \sin \phi$. The former will produce vibrations of frequency 2ω; the latter will give a mean force

$$\bar{F} = \frac{\omega^2 \bar{B}_{\parallel} B_{\perp} L A 2\pi r}{2(R^2 + \omega^2 L^2)}.$$

The quantities $B_{\parallel}$ and $B_{\perp}$ are related. An application of Gauss's theorem gives $2\pi r B_{\perp} \delta z = \pi r^2 \delta B_{\parallel}$, i.e. $\partial \bar{B}_{\parallel}/\partial z = 2B_{\perp}/r$ (z is the height). The expression for the force thus becomes

$$\frac{\omega^2 A^2 L}{4(R^2 + \omega^2 L^2)} \frac{\partial (\bar{B}_{\parallel})^2}{\partial z}.$$

The field $\bar{B}_{\parallel}$ will clearly decrease monotonically to zero as the height z increases, so that, provided the ring is not too heavy, there will be some point at which the calculated force is equal to the weight of the ring.

84

The calculations are straightforward, albeit providing ample scope for algebraic slips. The results are

(i) each magnet exerts a force on the other, of magnitude $3M_A M_B/x^4$, and in a direction perpendicular to the line AB. The two forces act in opposite senses;

(ii) each magnet exerts a couple on the other. These are not equal, and are both in the same sense. The couple acting on A is $2M_A M_B/x^3$, while that acting on B is $M_A M_B/x^3$. However, the two equal and opposite forces described in (i) constitute a third couple, of magnitude $3M_AM_B/x^3$, which is in the opposite sense to the other two. Thus the net force and the net couple acting on the whole system are both zero—as would be expected.

If follows that in any motion starting from rest, the centre of mass must not move and the total angular momentum about the centre of mass must remain zero. The two magnets will revolve about their common centre of mass, and each will rotate, in the opposite sense, about its centre. The couples are all of the same order of magnitude (M_AM_B/x^3), but the moment of inertia for the former motion is of order mx^2 (m is the mass of a magnet) while that for the latter is md^2 (d is the length of a magnet). The first motion will thus be very slow compared with the second. A will rotate twice as fast as B. The effect of the rotations is that M_A develops a component along AB, while M_B develops a component perpendicular to AB. The result is a force between the two acting along AB; consideration of particular cases shows this to be attractive—as would be expected. The details of the motion will depend on the relative values of x and d, but in the final state the two magnets will be at rest at their common centre of mass, lying side by side with unlike poles adjacent—the minimum-energy configuration.

85

The period τ is given by

$$\frac{\tau^2}{4\pi^2}\frac{3g}{2}=\frac{M_2l_2^2+M_1l_1^2}{M_2l_2-M_1l_1}$$

where M_1, M_2 are the two masses. If we denote the temperature by θ, then calculate $d\tau/d\theta$ and set it equal to zero, we eventually obtain

$$\alpha l^3M^2+(l^2+2l-2\alpha l^2-\alpha l)M-1=0 \tag{1}$$

where $M\equiv M_1/M_2$, $l\equiv l_1/l_2$ and $\alpha\equiv\alpha_1/\alpha_2$.

If we now write $M_1=\rho_1A_1l_1$, $M_2=\rho_2A_2l_2$, $\rho\equiv\rho_1/\rho_2$ and $A\equiv A_1/A_2=1$ according to the specification, then (1) can be rearranged to give

$$\alpha\rho^2l^5+\rho(1-2\alpha)l^3+\rho(2-\alpha)l^2-1=0. \tag{2}$$

The available materials restrict the range of possible values of α and ρ, but in principle l can take all values from 0 to $+\infty$. If we denote the expression in (2) by y, then $y\to+\infty$ as $l\to\infty$ (since ρ and α are both greater than zero for normal materials). For $l=0$,

$y = -1$. Thus there must be at least one real positive root of (2). This means that for any chosen values of α and ρ, a value of $l \equiv l_1/l_2$ which satisfies the given conditions can be found.

86

The time-averaged force, $(M+m)g$, exerted by the man-plus-ball system on the weighing machine will not be affected by interactions within the system, but only by interactions between the system and the outside world. Apart from possible minute effects due to air viscosity, these do not exist, and so the reading is $(M+m)g$.

A more detailed argument is as follows. The time of flight of the ball is $t = 2(2h/g)^{1/2}$, where h is the height risen. During this time, the mass on the platform is M. If the whole cycle occupies a time τ, then for a time $\tau - t$, the mass on the platform is $M+m$. Thus the time average of the weight would be $[Mgt + (M+m)g(\tau - t)]/\tau$. *But* the velocity of the ball on impact is $(2gh)^{1/2}$, and there is thus a downward impulse on the man of $m(2gh)^{1/2}$. There will be an equal impulse, also downwards, when he projects the ball upwards. Thus the total impulse per cycle is $2m(2gh)^{1/2}$; this is equivalent, when averaged over the cycle, to a force $2m(2gh)^{1/2}/\tau$, which must be added to the above expression. After simplifying and substituting $t = 2(2h/g)^{1/2}$, the result will be found to be $(M+m)g$, as before.

87

(*a*) Let us replace the bars by point masses at their centres. Then the attractive force between them is $G(\rho Al)^2/l^2$ (ρ is the density, l is the length, A is the cross-sectional area). The stress at the surface of contact is $GA\rho^2$. Assume that the mean compressive stress over the length of the bar is one half of this. Then the mean strain, and hence the total contraction, is easily found. It is equal to $G\rho^2 Al/2E$. (E is Young's modulus).

(*b*) The force between two elements with lengths dx, dy, situated at x, y, one in each bar, is $GA^2\rho^2\, dx\, dy/(x+y)^2$ (x and y are measured outwards from the junction). The total force on dx is found by integrating with respect to y from 0 to l. It is $G\rho^2 A^2 l\, dx/x(x+l)$. This will cause a stress in the length of bar between 0 and x of $G\rho^2 Al\, dx/x(x+l)$, and hence a contraction of this section by $G\rho^2 Al\, dx/E(x+l)$. This will be manifested as an inward movement of the end of the bar. The total movement due to the forces acting on all elements like dx is given by integrating with respect to x from 0 to l. It is $(G\rho^2 Al/E)\ln 2$.

The ratio of the two results is $2\ln 2 = 1.38$. (The contraction is of the order of 10^{-15} cm.)

88

The relative numbers of drops carrying a charge e, $2e$, $3e \ldots ne$ will depend on the conditions of the experiment, and the relative numbers chosen for measurement will also depend on the observer. It is likely that big drops will carry a larger charge than small drops; and while big drops will be more easily seen, and more likely to be measured on that account, a careful experimenter will know that small ones are more likely to give an accurate measurement. But when all such factors are taken into account, it is likely that the number measured which carry a charge ne will be a smooth function of n and, moreover, one which does not vary rapidly with n.

Hence if we notionally divide the drops into two groups carrying, respectively, odd and even numbers of charges, the population of the two groups will be approximately equal. Thus the probability that all the measurements actually made will be on members of one of the groups will be f^n, where $f \simeq \frac{1}{2}$ and n is the total number of measurements. If all drops carry an even number of charges, then $2, 4, 6 \ldots$ charges, each of value e, could be interpreted as $1, 2, 3 \ldots$ charges, each of value $2e$. The probability of getting an answer $2e$ from n observations is thus about $(\frac{1}{2})^n$. A similar argument shows that the chance of getting a wrong answer of $3e$ is $(\frac{1}{3})^n$—which is much smaller again, and can be neglected if $n > 20$. Thus the probability of getting a wrong answer of any size is about $(\frac{1}{2})^n$.

89

Let v be the velocity of water in the pipe at time t, increasing to $v + \delta v$ at $t + \delta t$. In this interval a mass $m = \rho A v \delta t$ has passed, where $A = \pi a^2$ is the cross-sectional area of the pipe. $M = \rho A h$ is the mass of water in the pipe. Equating the total energies at t and $t + \delta t$, we obtain

$$mgh + \tfrac{1}{2}Mgh + \tfrac{1}{2}Mv^2 = \tfrac{1}{2}Mgh + \tfrac{1}{2}M(v + \delta v)^2 + \tfrac{1}{2}mv^2.$$

This leads immediately to

$$\frac{dv}{dt} = \frac{2gh - v^2}{2h}$$

which integrates, with $v = 0$ at $t = 0$, to

$$v = (2gh)^{1/2} \tanh(g/2h)^{1/2}t.$$

Clearly $v \to (2gh)^{1/2}$ as $t \to \infty$.

With steady laminar viscous flow, the velocity v at radius r is $v_0(1 - r^2/a^2)$ where a is the radius of the pipe. The volume rate of flow is thus $V = \frac{1}{2}\pi v_0 a^2$ per second, and the pressure needed to maintain this flow in a horizontal pipe of length l is $p = 4\eta l v_0/a^2$ (Poiseuille flow). The work done per second is then $pV = 2\pi\eta l v_0^2$, and this energy is all dissipated as heat. In the case of the vertical pipe, the potential energy of the water leaving the tank per second can be equated to the sum of (a) the energy dissipated as heat, as calculated above, and (b) the kinetic energy of the outflow. This gives $\frac{1}{2}\pi v_0 a^2 \rho g h = 2\pi\eta h v_0^2 + \frac{1}{8}\pi\rho a^2 v_0^3$. This is a quadratic in v_0, and the one meaningful solution is $v_0 = (k^2 + 4gh)^{1/2} - k$, where $k = 8\eta h/\rho a^2$.

90

Two positive masses attract one another, so we must presume that a positive and a negative mass would repel one another, and therefore that two negative masses would attract. However, for a negative mass, force and acceleration will be in opposite directions.

(a) The force is attractive and both masses move towards one another.

(b) The force is again attractive, but both masses now move away from one another.

(c) The force is now repulsive. M_1, with the positive mass, will move away from M_2, but M_2 will move towards M_1. Since M_2 has the smaller mass it will have the larger acceleration and will eventually overtake M_1.

(d) The motions will be as in (c), but now M_1 will have the larger acceleration and will thus escape.

91

A particle, charge e, moving with velocity $\boldsymbol{v}$ in a field $\boldsymbol{B}$ experiences a force $e\boldsymbol{v} \times \boldsymbol{B}$ which, in the situation described, is directed radially inwards. If the particle is to move in an orbit of radius r, this force must be equal to mv^2/r, where m is the mass, or pv/r, where p is the momentum. Thus

$$pv/r = evB \tag{1}$$

and this remains true relativistically. If the velocity increases, the particle will continue to move in the same orbit, provided that B also increases to maintain the relation (1).

If the (varying) magnetic flux through the orbit is ϕ, there will be an EMF which, integrated around the orbit, is equal to $\mathrm{d}\phi/\mathrm{d}t$. A particle in the orbit will then also experience a tangential force $(e/2\pi r)\,\mathrm{d}\phi/\mathrm{d}t = F$, say. Its momentum at time t will thus be

$$p = \int_0^t F\,\mathrm{d}t = \frac{e}{2\pi r}\int_0^t \frac{\mathrm{d}\phi}{\mathrm{d}t}\,\mathrm{d}t = \frac{e\phi}{2\pi r}.$$

If this is to satisfy (1), we find, upon substitution, $\phi = 2\pi r^2 B$.

To investigate the stability, we write $B = B_0/r^n$ and suppose that the particle is originally in an orbit of radius r_0. Then, from (1), $pv/r_0 = evB_0/r_0^n$. If r_0 now increases to $r_0 + \delta r_0$, the inward radial force due to the field will become $evB_0/(r_0 + \delta r_0)^n$, while that needed to keep the particle in its new orbit is $pv/(r_0 + \delta r_0)$. The excess of the former over the latter, for small δr, is

$$\frac{evB_0}{r_0^n}\left(1 - \frac{n\delta r}{r_0}\right) - \frac{pv}{r_0}\left(1 - \frac{\delta r}{r}\right).$$

Using the above relation, this becomes $(evB_0/r_0^n)(\delta r/r_0)(1 - n)$ inwards. So, if $n < 1$, this force will act to reduce the orbit to its original size, i.e. the situation is stable against fluctuations of r. The possibility of oscillations of the orbit about its stable position must be considered in a full treatment.

92

Since the satellite drifts from west to east—i.e. in the same direction as the earth rotates—it is necessary to slow down its motion. Since it does 31 revolutions in 30 days we must increase its period by $1/30 = \delta\tau/\tau$.

One of the variants of the fundamental equation $GMm/r^2 = mv^2/r$ is $GM/4\pi^2 = r^3/\tau^2$—one of Kepler's laws. Thus if τ is to increase so also must r: the satellite must move into a higher orbit. Now the total energy of the body, $\frac{1}{2}mv^2 - GMm/r$, is, by virtue of the above relation, $-GMm/2r$. It thus follows that the total energy must be made to increase. If this change is brought about by a sudden change in velocity, so that the potential energy has no time to change, an increase in the total energy calls for an increase in velocity. Thus the force must be applied in such a direction that it accelerates the motion.

The immediate result will be that the new velocity is not related to the distance of the satellite from the earth in a way that is appropriate for a circular orbit. However, after the initial impulse, the satellite is moving freely, and its orbit must, in general, be an ellipse. This will have a common tangent with the original circle at the point that the impulse was applied (A). After half a revolution of the ellipse, it will again be moving in a direction normal to the radius vector and, if the impulse was correctly chosen, it will be at a distance from the earth equal to the desired radius of the enlarged orbit. However, its velocity will not now be of the correct value to permit it to move on this desired circular orbit. It must therefore be given a second impulse to bring the velocity to the correct value. Again, the ellipse and the larger circle will have a common tangent at the point where the impulse was given (B), and it can be seen from the diagram that this impulse, too, must be in a sense to accelerate the motion, and of a suitable magnitude.

It is not necessary, for the present purposes, to investigate the details of the intermediate stages. The net effect has been to move the satellite from one circular orbit to another, larger, circular orbit and thus its velocity must have been reduced—in spite of the fact that the immediate effect of both impulses was to increase the velocity. Another variant of the fundamental equation is $\tau = 2\pi GM/v^3$; hence we see that $\delta\tau/\tau = -3\delta v/v$. In the present problem we must therefore have $\delta v/v = -\frac{1}{90}$. To obtain a numerical value we may note that $GM/R_0^2 = g$, where R_0 is the

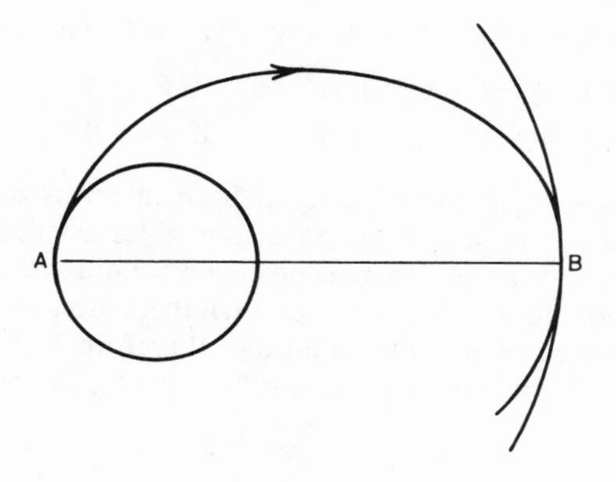

radius of the earth. This gives $v^3 = 2\pi g R_0^2/\tau$. Since the satellite is in a geostationary orbit, $\tau = 24$ hours. This gives $v = 3080\ \mathrm{m\,s^{-1}}$, and hence $\delta v = 34\ \mathrm{m\,s^{-1}}$.

The above argument has been spelled out in detail, since the apparent paradox can be very confusing.

(By way of a bonus, we can consider the motion around the intermediate, semi-elliptical path. Both angular momentum and *total* energy must be conserved, and by equating the values at the two ends we can obtain expressions for the initial and final velocities. These are $(2GMr_2/r_1(r_1+r_2))^{1/2}$ and $[2GMr_1/r_2(r_1+r_2)]^{1/2}$ respectively. Since the velocities before the first and after the second impulse were $(GM/r_1)^{1/2}$ and $(GM/r_2)^{1/2}$, we find that at each impulse, the kinetic energy increases by $\frac{1}{4}GMm\delta r/r^2$. When combined with the change in the gravitational energy, $-GMm\ \delta r/r^2$, this gives the correct change in total energy.)

93

Let the waves be travelling in the positive x direction. Their wavelength is $\lambda = v\tau$. If the ship's course makes an angle θ with the x axis, then the component of the ship's velocity along Ox is $V\cos\theta$, and the relative velocity of ship and wave is $V_r = V\cos\theta - v$. V_r may have either sign, depending on whether the ship overtakes the waves or vice versa.

The time for the ship to travel from crest to crest is

$$t = \left|\frac{\lambda}{V\cos\theta - v}\right| = \left|\frac{v\tau}{V\cos\theta - v}\right|$$

and the condition for a sensitive course is $t = T$. This gives

$$\cos\theta_1 = (1 + \tau/T)v/V \qquad \text{if } V_r > 0$$
$$\cos\theta_2 = (1 - \tau/T)v/V \qquad \text{if } V_r < 0.$$

The quantities τ, T, v and V have all been taken as positive, so $\cos\theta_2 < \cos\theta_1$. If either of the equations gives a value of $\cos\theta$ outside the range ± 1, then there is no corresponding real value for θ. If the solution lies inside this range, then there are two solutions, since $\cos(-\theta) = \cos\theta$. For *four* solutions, therefore, both values of $\cos\theta$ must lie within ± 1. Because $\cos\theta_2 < \cos\theta_1$, this means that

$$\cos\theta_1 \leqslant 1 \qquad \cos\theta_2 \geqslant -1$$

i.e.

$$V/v \geqslant 1 + \tau/T \qquad -V/v \leqslant 1 - \tau/T$$

or

$$V/v \geqslant \tau/T + 1 \qquad V/v \geqslant \tau/T - 1.$$

If the former condition is satisfied, then so also is the latter, so the condition for four sensitive courses is

$$V/v \geqslant \tau/T + 1.$$

(The above treatment regards the ship as a point. If its length is L, then the number of waves interacting with the ship at any one time is $L \cos \theta/\lambda = n$, say. The effect on the ship may be taken to be proportional to the mean wave height averaged over the length of the ship. This is readily found by a simple integration; it is equal to $2\pi a(\sin n\pi)/n\pi$, where $n = L \cos \theta/\lambda$. This is a maximum for $L = 0$, i.e. the 'point' ship, and oscillates as L increases, approaching zero as $L \to \infty$. It has zeros at $L \cos \theta/\lambda =$ an integer, which means that some of the sensitive courses may not be sensitive at all, if the ship happens to be the right length.)

94

Let x, y be the positions of the monkey and the bananas, measured upwards, and let T, S be the tensions in the cord which supports the monkey and the bananas. Then the three equations of motion are

$$m\ddot{x} = T - mg \qquad m\ddot{y} = S - mg \qquad (T - S)a = I\dot{\omega}$$

where a is the radius and I is the moment of inertia of the pulley. Eliminating $T - S$ and integrating once, we obtain $m(\dot{x} - \dot{y}) = I\omega/a$. But if the cord is of fixed length the motion of the bananas is related to that of the pulley by $\dot{y} = a\omega$. Making this substitution and rearranging, we obtain $\dot{x}/\dot{y} = (ma^2 + I)/ma^2$, i.e. $\dot{x} > \dot{y}$: the monkey overhauls the bananas. This is case (i) of the addendum. If $I = 0$, we have the original problem and $\dot{x} = \dot{y}$, i.e. the separation is constant.

In any pendulum, the mean value of the tension in the string, when swinging, is greater than its static value by the amount needed to cause the mass to move on its circular path. Thus the effective mass of the monkey is increased by his movements, i.e. he will lose ground.

95

The period of a compound pendulum is

$$\tau_0 = 2\pi(k^2/lg)^{1/2} = 2\pi(L/g)^{1/2}$$

where k is the radius of gyration about the pivot, l is the distance of the centre of mass from the pivot and L is the length of the 'equivalent simple pendulum'. If a small mass m is placed at a distance x from the pivot, the period becomes

$$\tau = \frac{2\pi}{g^{1/2}}\left(\frac{Mk^2 + mx^2}{Ml + mx}\right)^{1/2}$$

where M is the mass of the pendulum. The most effective position is found by maximising $(Mk^2 + mx^2)/(Ml + mx)$ with respect to x. The result is $x = k^2/2l = L/2$, where the approximation $mk^2/Ml^2 \ll 1$ has been made.

If this is substituted in the equation for τ, we obtain

$$\left(\frac{\tau}{\tau_0}\right)^2 = \frac{1+Z}{1+2Z}$$

where $Z \equiv mk^2/4Ml^2$ and $\tau = \tau_0$ when $m = 0$. Since Z must be positive, $\tau < \tau_0$, i.e. the clock is speeded up by adding m. If we write $\tau = \tau_0(1-\epsilon)$ and approximate, we get $Z = 2\epsilon$. An error of 4 minutes per day corresponds to $\epsilon = 1/360$. For the simplest type of pendulum (a heavy bob on a light rod), $k \simeq l$. Thus $m/M = 0.022$ and $m = 22$ g—hardly a 'small weight'!

96

We draw up a table giving the 'dimensions' in length and time of the seven quantities listed, and of the operator ∇. Since we are interested in space inversion, only those lengths which are vector quantities should be considered. The dimensions of $\boldsymbol{s}$ follow from its definition as an angular momentum, $\boldsymbol{s} = \boldsymbol{r} \times \boldsymbol{p}$; those of $\boldsymbol{E}$ from the equation for the field of a charge, i.e. $\boldsymbol{E} \propto \boldsymbol{r}/|\boldsymbol{r}|^3$; and those of $\boldsymbol{B}$ from the relation $\partial \boldsymbol{B}/\partial t \propto \nabla \times \boldsymbol{E}$. Then, if any quantity has an odd number of dimensions in length or time, it will change sign under P or T, respectively. In the table, a quantity which changes sign is denoted by $-$; one which does not is denoted by $+$.

Quantity	r	t	v	p	S	E	B	∇
Dimensions	L	T	LT^{-1}	LT^{-1}	L^2T^{-1}	L	T	L^{-1}
Under P	−	+	−	−	+	−	+	−
Under T	+	−	−	−	−	+	−	+

We now carry through a similar process for each of the equations, considering each term in turn, and using the ordinary laws of algebra (the product of two negative quantities is positive). If, under any transformation, the result for all the terms in an equation is the same, then the equation is invariant under that transformation.

Equation (a)

Term	p/t	E	vB	Result
Under P	− +	−	− +	Invariant
Under T	− −	+	− −	Invariant

Equation (b)

Term	rt^{-2}	r	Result
Under P	− + +	−	Invariant
Under T	+ − −	+	Invariant

Equation (c)

Term	sv	Constant	Result
Under P	+ −	+	Not invariant
Under T	− −	+	Invariant

Equation (d)

Term	Scalar	sB	Result
Under P	+	+ +	Invariant
Under T	+	− −	Invariant

97

The frequency of the lowest mode of vibration will be of order $\nu = V/2l$, where V is the velocity of sound and l is the linear dimension of the solid, and the corresponding energy is $h\nu$. Only this least energetic mode need be considered. The probability that it will be excited is $\exp(-h\nu/k_{\mathrm{B}}T)$. Strictly, we should use the Bose–Einstein formula, but under the conditions of this problem the difference is negligible. If we set $\exp(h\nu/k_{\mathrm{B}}T)$ equal to some large number, say 10^4, we obtain the condition that the particle should be above absolute zero for 10^{-4} of the time, i.e. it should be at zero for 99.99% of the time. The condition is thus $hV/2lk_{\mathrm{B}}T = \log_e 10^4$. The velocity of sound in a solid may be taken to be typically 3×10^3 m s^{-1}; this gives $l \simeq 0.8$ mm.

98

We enquire into the phases of the contributions from the two slits to the resultant vibration at a general point P on the screen. Let P′ be the perpendicular projection of P on to the plate containing the slits, and x, y, the coordinates of P′, as shown. Then, provided x and y are sufficiently small that, by comparison, the slits can be regarded as being infinitely long, the phase of the resultant vibration at P due to the separate contribution from each slit will be determined only by the perpendicular distance to this slit from P. The phase difference between the contributions from the two slits will be equal to

$$[(l^2 + x^2)^{1/2} - (l^2 + y^2)^{1/2}]2\pi/\lambda.$$

The fringes are the loci of points for which this phase difference is a constant. Using a binomial expansion, this condition can be written as $x^2 - y^2 = \text{constant}$. The fringes are thus rectangular hyperbolae, as shown.

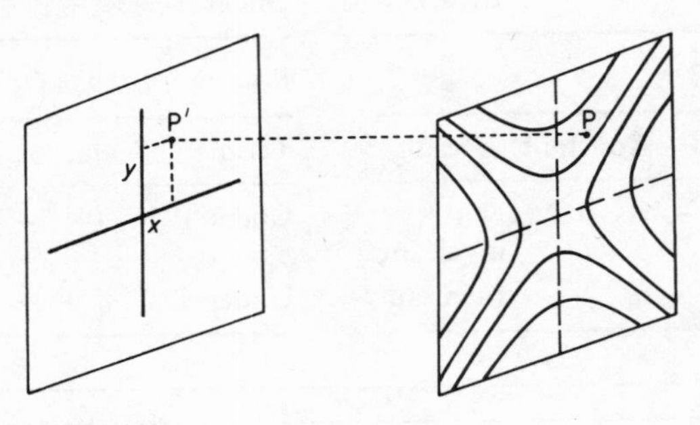

99

The linear dimensions of a neutron (nucleus) are smaller than those of an atom by a factor of about 10^{-5}. The suggested collapse would therefore reduce the radius of the sun by a factor $f \simeq 10^{-5}$. This assumes that the particles in both the sun and the neutron star are 'touching'. The data given allow this to be checked for the sun. Its density comes out to be $1.3 \times 10^3\ \text{kg m}^{-3}$—about the same as that of water. The assumption is also reasonable for a neutron star.

If the radius changes by a factor f, while the mass is constant, the moment of inertia $I\ (\propto mr^2)$ will change by a factor f^2. If the angular momentum $(= I\omega)$ is constant, ω will change by f^{-2}, and hence the period ($\propto 1/\omega$) will change by f^2. If we assume that we can treat the sun as a uniformly magnetised sphere, with intensity of magnetisation J, then the flux through the equatorial plane is proportional to Jr^2, and the magnetic moment M is proportional to Jr^3. If r changes by a factor f, while the flux stays constant, J will change by f^{-2}, and M by $f^{-2} \times f^3 = f$. The field at large distances is proportional to M, and this too will change by f. The field on the surface will be the dipole field due to $M(= M/r^3)$, and will therefore change by $f/f^3 = f^{-2}$.

If $f = 10^{-5}$, the period falls by a factor 10^{10}, the surface field rises by a factor 10^{10} and the remote field falls by a factor 10^5.

100

The field between the electrodes is $E = V/r \ln(r_2/r_1)$. This will induce a dipole $m = 4\pi\epsilon_0 a^3 E$ on a spherical particle of radius a, and the particle will thus experience a radial force

$$m \frac{\partial E}{\partial r} = \frac{4\pi\epsilon_0 a^3 V^2}{r^3 (\ln r_2/r_1)^2}.$$

There will also be the axial force due to gravity, $\frac{4}{3}\pi a^3 \rho g$. From the values given we find that the former is of the same order as the latter. (It is, in fact, about 20 times as big close to the central electrode, and thus somewhat smaller over most of the volume.) Since the particles take 'several minutes' to settle, they are clearly small enough for Stokes' Law to apply, i.e. force $\propto$ velocity and not acceleration. The same will therefore be true for the radial motion. We can then write the axial velocity as $v_z = Kmg$ and the radial velocity as $v_r = Kb/r^3$, where K is the same constant in both expressions and $b \equiv 4\pi\epsilon_0 a^3 V^2/(\ln r_2/r_1)^2$. The second of these relations shows that the time taken by a particle to reach the central electrode, starting from a radius r, is approximately $r^4/4Kb = t$, say. During this time it will have moved a vertical distance $Kmgt = mgr^4/4b = z_0$. If it started at a distance from the bottom less than z_0, it will miss the central electrode. Thus all those particles within a cylindrical annulus of radius r and height z_0 will escape, i.e. a total number

$$\int_{r_1}^{r_2} 2\pi r \, dr \, z_0 n$$

where n is the number per unit volume. With the above expression for z_0, this becomes $\pi mgnr_2^4/12b$. The total number present in the cylinder is $\pi r_2^2 ln$ (l is the length). Collecting these results, we find that the fraction which miss the central electrode is

$$\frac{\rho g r_2^4 (\ln r_2/r_1)^2}{36\epsilon_0 l V^2} \simeq 0.2.$$

Note that the errors introduced by neglecting higher powers of r_1/r_2 are quite small. However, the above argument ignores the 'end effects' in the electric field; in the present context these are clearly very important, with the result that the final answer may be quite wrong.

101

The geometry of this effect is dominated by the fact that the line of sight from the observer meets the water surface at almost grazing incidence. Consequently only the fronts of the crests of the waves take part in the reflection. We first consider the optics in a vertical plane containing the moon and the observer, and we idealise the sea surface as a sine wave of constant amplitude a and wavelength λ, with the line of the crests normal to this vertical plane. Unless the sea is rather rough, a/λ will not be larger than about 0.04, corresponding to a maximum slope of the wave surface of about 15°. The relevant line of sight, which makes an angle α with the horizontal ($\alpha < 1/10$ rad), grazes the top of one wave and is shown in the diagram meeting the next wave at P. The diagram has a greatly exaggerated vertical scale.

If the wave profile is $y = a\cos 2\pi x/\lambda$, and if $z \equiv a - y$, then, referring to the diagram,

$$z = a(1 - \cos 2\pi x/\lambda) \simeq \tfrac{1}{2}a(2\pi x/\lambda)^2 \qquad (1)$$

and

$$\alpha = z/(\lambda - x) \simeq z/\lambda = 2\pi^2 ax^2/\lambda^3. \qquad (2)$$

The slope of the wave surface at P is $(2\pi a/\lambda)\sin 2\pi x/\lambda = 4\pi^2 ax/\lambda^2 = \gamma$, say. The angle β between the reflected line of sight and the horizontal marks the limit of that part of the sky that can be seen by reflection in some part of the wave surface. Clearly

$$\beta = 2\gamma + \alpha = (8\pi^2 ax/\lambda^2) + \alpha$$

i.e., from (2),

$$\beta = 4\pi(2a\alpha/\lambda)^{1/2} + \alpha$$

or

$$\alpha = \frac{\lambda}{2a}\left(\frac{\beta - \alpha}{4\pi}\right)^2. \tag{3}$$

With $a/\lambda = 0.04$, this means that if the elevation of the moon above the horizon (β) is anything less than 30°, then any reflected light seen by the observer will lie within an angle α below the horizon, and $\alpha \simeq 1^\circ$.

It remains to consider the consequences of the adoption of a highly idealised model of the real situation. (i) The waves are not sinusoidal; their troughs are flatter, and their crests more curved. Since we have approximated the shape of the crest by a parabola (equation(2)), this will still be a reasonable approximation for a real wave, but with an effective value of a/λ bigger than if it had been a sine wave. (ii) The waves are not all the same size. The small waves, hiding in the troughs between the large ones, might just as well not be there at all, as far as contributions to the reflection are concerned. Thus the effective value of λ may be several times the real wavelength, and this effect will be more important than the fact that the effective value of a will correspond to the largest waves: thus a/λ will be decreased as a result. (iii) The lines of the crests will not all be normal to the line of sight. Those which are nearly normal will contribute in a similar way to that already described, but will give rise to reflections outside the vertical plane considered hitherto, i.e. the blaze of light will spread sideways.

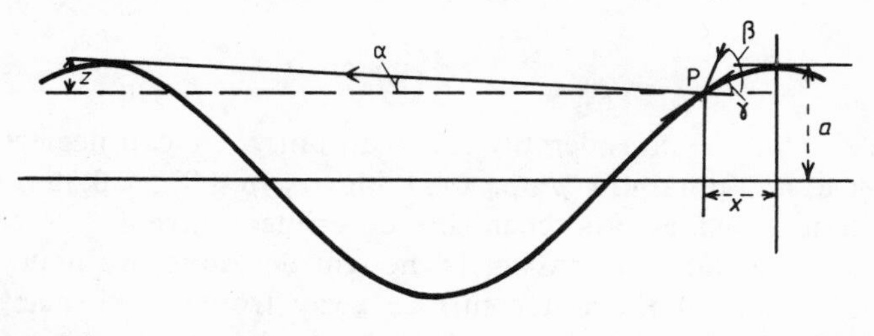

102

Several approximations follow from the fact that the clearance is very small compared with the diameter. The upward velocity of the liquid will be large compared with the downward velocity of the slug. We can thus treat the liquid flow as if between two large, stationary, flat plates, separated by $2d$. Standard methods give the velocity profile

$$V = V_0(1 - x^2/d^2).$$

(a) The average flow velocity is then found to be $\frac{2}{3}V_0$. The volume rate of flow of liquid upwards is thus $\frac{2}{3}V_0 2\pi r 2d$, where r is the radius of the slug (or tube); this is equal to the rate of displacement of liquid by the slug, $\pi r^2 v$ (v is the velocity of the slug).

(b) The velocity gradient in the liquid, $\partial V/\partial x$, is equal to $2V_0/d$ at $x = d$, and thus the viscous drag on the slug is $2\pi r l \eta 2V_0/d$, where l is the length of the slug. This is balanced, in the steady state, by the weight of the slug less its buoyancy, i.e. $\pi r^2 l(\rho - \rho_0)g$.

If we eliminate V_0 between the two equations resulting from (a) and (b), we find $v = 2(\rho - \rho_0)gd^2/3\eta$, which is independent of r or l!

103

Let A denote the area, L the perimeter and ρ_m the density of a blade; let ρ be the density and γ the surface tension of water. Then for equilibrium we have

$$\rho_m g A t = \rho g h A + \gamma L \sin\alpha$$

where t, h and α are as shown in diagram (a). For an order-of-magnitude argument, we can write $A = a^2$ and $L = 4a$. Then, if we also write $h = (\gamma/\rho g)^{1/2}$ as suggested in the problem, we obtain, on dividing by $\rho g a A$,

$$\frac{\rho_m}{\rho}\frac{t}{a} = \frac{h}{a} + 4\left(\frac{h}{a}\right)^2 \sin\alpha.$$

Since h/a will be considerably less than unity, we can neglect the last term, and obtain $t = \rho h/\rho_m$ or, taking $\rho_m/\rho \simeq 8$, $t = 0.35$ mm—i.e. about $3\frac{1}{2}$ blades (a kitchen-sink experiment gave 4).

(If a more elaborate answer is thought desirable, we may note that the shape of the water surface away from the corners can be found, without approximation, from the familiar equation $\Delta p = \gamma(1/r_1 + 1/r_2)$. In the present case, $\Delta p = \rho g z$, where z is the distance below the surface, $r_1 = \infty$ and $1/r_2 = \mathrm{d}\theta/\mathrm{d}S$, where θ is the inclination of the surface profile to the horizontal and S is the distance measured along the arc from an arbitrary zero. Now $dz/ds = \sin\theta$, so we obtain

$$d\theta/dz = \rho g z/\gamma \sin\theta$$

which can be integrated at once to give

$$\rho g z^2/2\gamma = 1 - \cos\theta$$

since $\theta = 0$ at $z = 0$. This can be written

$$(z/h)^2 = 2(1 - \cos\theta)$$

where h is the 'surface tension depth' as defined. It is obvious that the maximum possible value of z, corresponding to $\cos\theta = -1$, is $z = 2h$. The shape of the curve is shown (approximately) in part (*b*) of the diagram. It is doubtful whether values of θ greater than $\pi/2$ would be realisable in practice; this would give $z = \sqrt{2}h$, which is approximately equal to h, as suggested in the problem.)

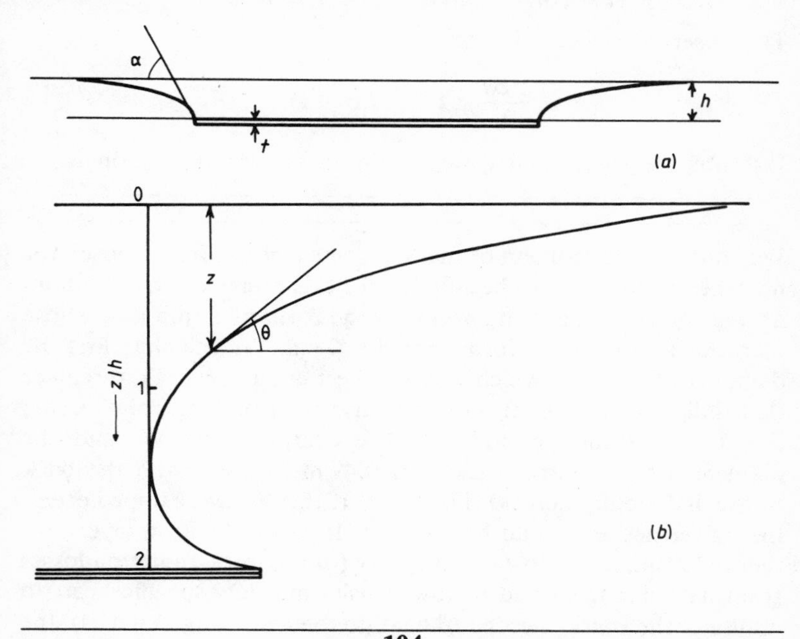

104

The velocity of light in the medium is c/n, where n is the refractive index. If the medium is moving with a velocity v parallel to the direction of light propagation, then the velocity as seen by a stationary observer is given by the Lorentz transformation

$$v_x = \frac{v_x' + v}{1 + vv_x'/c^2}$$

i.e.

$$V = \frac{c/n + v}{1 + v/nc}.$$

For $v \ll c$, this gives

$$V \simeq c/n + (1 - 1/n^2)v.$$

In the Fizeau-type experiment described, the time difference for the two light beams is

$$\Delta t = \frac{2l}{(c/n) - (1 - 1/n^2)v} - \frac{2l}{(c/n) + (1 - 1/n^2)v} \simeq \frac{4n^2 vl}{c^2}\left(1 - \frac{1}{n^2}\right).$$

The observed fringe shift is

$$\frac{c\,\Delta t}{\lambda} = 4\,\frac{v}{c}\,\frac{l}{\lambda}\,(n^2 - 1).$$

Inserting the values given, we obtain an answer of 4.1 fringes.

105

We consider the problem of melting some piece of iron, chosen for no other reason than to be suitable for this exercise. The optimum arrangement will be if the iron is in the form of a thin disc whose diameter is that of the focal spot, i.e. 2 cm. We calculate first the diameter of a mirror which would collect enough energy to replace that radiated from the front of the disc at its melting point. Using Stefan's constant we find that the energy is 190 W, and the diameter of the mirror needed is 0.49 m. Losses from the back of the disc would increase this figure. If the disc were unprotected, the power needed would be doubled. If covered with a layer of a thermal insulator of low conductivity (to minimise radiation losses from its outer face) and of low density and low specific heat (to minimise the energy needed to heat up the insulating material), the power increase would be quite small. The above assumes that the mirror is a perfect reflector and the iron a perfect absorber. While the former may be almost true, the latter will not. A knowledge of the absorption coefficient over the spectral and temperature ranges involved would permit a correction to be made. Typical values range from about 0.3 for a light surface to 0.8 for a dark one, so that this factor might well double the power required. Convective losses have also been ignored: these would not be negligible, but are difficult to calculate.

In the early stages of the heating, the heat losses would be very small, and almost the whole of the power input would be available for raising the temperature. If the disc is 1 mm thick, its mass is about 2.5 g and its thermal capacity is about 1.1 $\mathrm{J\,K^{-1}}$. Thus the incident 190 W—if all absorbed—would raise the temperature at about 170 $\mathrm{K\,s^{-1}}$. This rate would soon drop, as the losses became important, and the melting point would only be approached asymptotically. But a 10% increase in the power input would produce a 2.5% increase in this asymptote—i.e. about 40 K—and by the time the melting point was reached, the temperature would still be rising at about 17 $\mathrm{K\,s^{-1}}$. To produce melting, the latent heat would also have to be supplied. An estimate of the value indicates that about 600 J would be needed: the 10% surplus power could supply this in half a minute. Altogether, the process might take a minute from the start.

The general conclusion is that while the calculated minimum size of 0.5 m diameter would certainly not be adequate, it would be worthwhile trying with a mirror twice as big, with some hope of success in a time of the order of one minute.

106

This problem calls for a solution of the heat conduction equation with a moving-boundary condition. This is difficult, but since the changes will be slow it will be a good approximation to assume that the temperature varies linearly through the ice at all times. Thus we can write

$$\theta = \theta_0 + \Delta\theta x/d$$

where $\theta_0 = -10^\circ\mathrm{C}$, $\Delta\theta = 10^\circ\mathrm{C}$, d is the thickness of the ice and x is the distance below the top surface. We write down an equation expressing the fact that the heat flow out through the top surface in time δt is sufficient to freeze an additional layer of water of thickness δd, and to cool the existing ice by an amount $\delta\theta$—which may be a function of both position and time. Referring all quantities to unit area, we obtain

$$\varkappa\left(\frac{\partial\theta}{\partial x}\right)_{x=0}\delta t=\rho L\ \delta d-\int_0^d \rho\sigma\ \delta\theta\ \mathrm{d}x.$$

From the previous equation, we have $\delta\theta=\Delta\theta x\ \delta d/d^2$, and the integral thus has the value $-\frac{1}{2}\rho\sigma\ \Delta\theta\ \delta d$. We also have $\partial\theta/\partial x=\Delta\theta/d$. These substitutions give

$$\varkappa\frac{\Delta\theta}{d}=\rho L\frac{\delta d}{\delta t}+\frac{\rho\sigma\ \Delta\theta}{2}\frac{\delta d}{\delta t}$$

i.e.

$$\frac{\partial d}{\partial t}=\frac{\varkappa\ \Delta\theta}{\rho d(L+\sigma\ \Delta\theta/2)}$$

or

$$d^2=\frac{2\varkappa\ \Delta\theta t}{\rho(L+\sigma A\theta/2)}.$$

If the effect of specific heat is to be neglected, we can write $\sigma=0$. Inserting the numerical values, we obtain $d=0.72$ m if the effect of specific heat is ignored, and 0.71 m if it is not.

107

A conductor carrying a surface charge density σ will experience an outward normal stress $\sigma^2/2\epsilon\epsilon_0$. For a soap bubble in air, this will be equivalent to an internal pressure $\sigma^2/2\epsilon_0$.

The pressure in the original bubble is $p+4\gamma/r$, where p is the atmospheric pressure. When it doubles its radius, this will be reduced by a factor of 8. The extra effective pressure due to the electrostatic forces will then be needed to produce equilibrium against the new surface tension force $4\gamma/2r$. Thus

$$\tfrac{1}{8}(p+4\gamma/r)+\sigma^2/2\epsilon_0=p+2\gamma/r.$$

This equation gives σ. The total charge on the enlarged bubble is $16\pi r^2\sigma$, which leads directly to the result quoted.

108

Let θ denote the vertex angle of each cone, D the diameter of the lens and r the distance of the photosensitive cell from the vertex of the cone. The focal length of the lens is then $R-r$, and the radius of the Airy disc formed in its focal plane is $1.22\lambda(R-r)/D$. For maximum sensitivity, the diameter of the sensitive cell ($\equiv d = rD/R$) should equal the diameter of the Airy disc, i.e.

$$rD/R = 2.44\lambda(R-r)/D. \tag{1}$$

For good angular resolution, the fields of view of adjacent cones should not be confused. Since the angular resolution is $\theta = D/R$, this means that

$$D/R = \theta = 1.22\lambda/D. \tag{2}$$

Thus, from (1) and (2), $rD/R = 2(R-r)1.22\lambda/D = 2(R-r)D/R$ and hence

$$r = 2R/3.$$

From equation (2), $D^2 = 1.22\lambda R$ and $\theta = D/R$. For $\lambda = 500$ nm and $R = 1$ mm we find $D = 2.5 \times 10^{-5}$ m and $\theta = 0.027$ rad $\simeq 1.6°$.

109

For the alloy the mean atomic weight is $(0.95 \times 55.85) + (0.05 \times 10.82) = 53.60$ $(= \overline{M})$. The spacing of [110] planes in a BCC crystal is $a/\sqrt{2}$, where a is the interatomic distance. This same quantity is given by Bragg's law as $\lambda/2 \sin\theta$. Thus from the data given we find $a = 2.874 \times 10^{-10}$ m. If the B atoms are in substitutional positions, then, since there are two atoms per unit (cubic) cell in the BCC structure, the density calculated from x-ray data will be $2\overline{M}/a^3 Z$ (Z is Avogadro's number), which is equal to 7.495×10^3 kg m^{-3}. Since this is much less than the measured bulk density (7.85×10^3), there must be extra mass somewhere in the unit cell—i.e. the B atoms must be on interstitial sites.

The density of pure A can be used to check that the bulk density agrees with the value obtained from x-ray data in this case: it does. The values of e and h are clearly irrelevant.

110

Elementary dimensional analysis gives the drag force $D = C_D \rho A v^2$. Let α be the (constant) acceleration. Then $v = \alpha t$ and $h = \frac{1}{2}\alpha t^2$; thus

$$D = C_D \rho_0 \exp(-h/h_0) \qquad A\alpha^2 t^2 = K t^2 \exp(-\alpha t^2/2h_0)$$

where K is a constant. Setting $\mathrm{d}D/\mathrm{d}t = 0$ for the maximum, we obtain $t_{max}^2 = 2h_0/\alpha$ and hence $h_{max} = h_0$.

111

If $F = F(t)$ is the frictional force exerted by the ground on the hoop, we have

$$v = v_0 - \int F/m \, dt \qquad \omega = \omega_0 - \int Fr/I \, dt.$$

By the time that slipping between the hoop and the ground has stopped, we have $v + r\omega = 0$, i.e.

$$v_0 + r\omega_0 = (1/m + r^2/I) \int F \, dt.$$

By this time the velocity of the centre of mass is $v_0 - (1/m)\int F \, dt$. Inserting the value of $\int F \, dt$ from the previous equation and writing in the requirement that v is to be negative, we obtain the condition

$$I\omega_0 > mrv_0.$$

For a hoop, $I = mr^2$, giving $\omega_0 > v_0/r$, as required. For a disc, $I = \frac{1}{2}mr^2$, giving $\omega_0 > 2v_0/r$.

112

The gradient of the moon's gravitational field, in the neighbourhood of the earth, has the effect of raising two bulges on the surface of the oceans, as shown, grossly exaggerated, in the diagram. The bulges always point towards and away from the moon, respectively. As the earth rotates, a wave with two crests and two troughs propagates around the earth once every 24 hours. The maximum height of the bulges clearly occurs at latitude $\pm 23°$, on account of the inclination of the axis mentioned in the question. Thus point A in the diagram, which is having a high tide now, will be at B 12 hours later, where it will again have a high tide—but of a different height. The size of this difference can be calculated thus.

We assume that the solid earth is a sphere, and that the surface of the water is an ellipsoid of revolution. In polar coordinates, the equation of an ellipse is, approximately, $r = b(1 + \delta \cos^2 \phi)$. If this is assumed to describe the water surface, then the surface of the solid earth will be given by $r = b'$, where $b' \simeq b$, i.e. the depth of water is approximately $b\delta \cos^2 \phi$. If the points A and B in the diagram are at a latitude θ then, for A, $\phi = (\theta + 23°)$ and, for B, $\phi = (\theta - 23°)$. The difference between the heights of the two high tides is thus $b\delta[\cos^2(\theta - 23°) - \cos^2(\theta + 23°]$, which can be transformed into $b\delta \sin 2\theta \sin 46°$. Thus the difference between high tides is zero on the equator and at the poles and a maximum at latitude 45°.

As the moon revolves about the earth there will be two occasions every month when the earth–moon line coincides with the projection of the earth's axis on the plane of the moon's orbit. This is the situation shown in the diagram. The difference in height of successive high tides will be a maximum at these times, and will fall to zero half-way between.

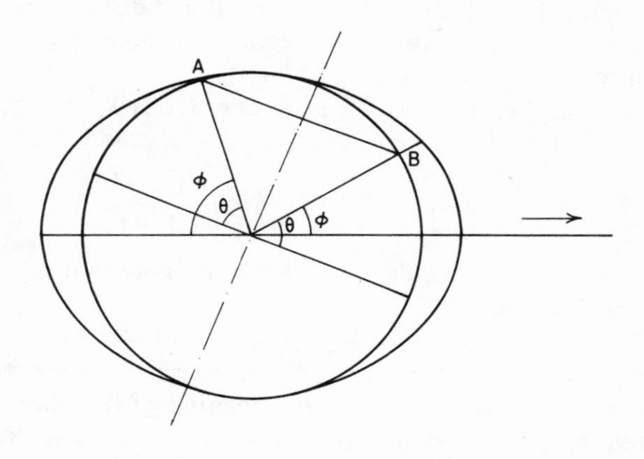

113

Let the radius of the circular cross section of the film at a height z above the lower ring be $r(z)$. To support the ring, of weight w, we must have (resolving vertically at height z)

$$2\pi r 2\gamma[1 + (\mathrm{d}r/\mathrm{d}z)^2]^{-1/2} = w$$

where γ is the surface tension. Thus

$$\frac{\mathrm{d}r}{\mathrm{d}z} = \left[\left(\frac{4\pi\gamma}{w}\right)^2 r^2 - 1\right]^{1/2}$$

and so

$$z = (1/c)\ln[cr + (c^2r^2 - 1)^{1/2}] + K$$

where K is a constant. Here the formula given in the question has been used, and $c \equiv 4\pi\gamma/w$. The difference in height between the rings is thus

$$\frac{1}{c}\ln\left(\frac{3ac+(9c^2a^2-1)^{1/2}}{ac+(c^2a^2-1)^{1/2}}\right).$$

The maximum distance at which the ring can hang in stable equilibrium is derived by setting $dz/dc=\infty$, which happens at $c=1/a$. The distance then is $a\ln(3+\sqrt{8})=1.76a$.

114

If the distance moved at any stage is r, and a step of length s is then made in an arbitrary direction, the resultant distance is t, where $t^2=r^2+s^2-2rs\cos\theta$. Since all directions are equally probable, we find, on taking the average, $\overline{t^2}=r^2+s^2$.

In this case, the RMS distance after one step is λ, after two steps, $\lambda(1+f^2)^{1/2}$, after three steps, $\lambda(1+f^2+f^4)^{1/2}$ and, after N steps,

$$(1+f^2+\cdots+f^{2(N-1)})^{1/2}=\lambda\left(\frac{1-f^{2N}}{1-f^2}\right)^{1/2}$$

which is the required result. As a check, we note that if $f\rightarrow 1$, this expression becomes $\lambda N^{1/2}$.

115

It is necessary to be clear about the meaning of the phrase 'widely separated fringes'. The fringe separation is $x=\lambda D/d$ where d is the slit separation and D is the distance from slits to screen (eye). If an eye is to be placed 'at' a bright fringe, it means that x must be a good deal larger than the diameter of the pupil, p. We write $x=np$, where n is of the order of 10, say. Then $\lambda D/npd\simeq 1$.

The angular resolving power of the eye is approximately λ/p, and the angular separation of the slits as seen by the eye is d/D. If the slits are to be resolved, we must have $d/D>\lambda/p$, i.e. $\lambda D/pd<1$. But we have just shown that this quantity is equal to n. Thus if the conditions as set out are satisfied, the eye will not be able to resolve the slits. If placed at a bright fringe, then, depending on the relation between fringe spacing and pupil diameter, the eye will see a single broad and fuzzy slit or an almost uniform distribution of intensity. This will be four times the value that would be given by a single slit. With the eye placed at a dark fringe, it will detect almost nothing—but not quite, since it is not possible for the whole pupil to be 'at' the centre of the fringe.

(The use of the concept of angular resolution is not strictly valid in this context, since it is based on the distribution of intensity from two neighbouring incoherent sources. A full treatment for the two coherent sources of this problem is much more difficult, but the analogous situation in the Abbe theory of microscope resolving power may be considered. It is shown there that if one is to obtain any significant resolution of a periodic object, at least two orders of diffraction must enter the objective. In the present problem one order (one fringe) is used, so we might again expect that the two slits would not be resolved.)

116

The neutron wavefunctions are approximately plane waves. Initially, both ψ_1 and ψ_2 are proportional to $\exp(-k_1x)$, with $k_1^2 = 2mE/\hbar^2$, where m is the neutron mass and E is its kinetic energy. If beam 2 is displaced downwards by a distance d, the potential energy of a neutron is decreased by mgd, and hence $\psi_2 \propto \exp(ik_2x)$, with $k_2^2 = 2m(E + mgd)/\hbar^2$. Beam 1 is not disturbed. The change in k will be small, so we can write $k_2 = k_1 + \Delta k$ and neglect terms in $(\Delta k)^2$. This gives $\Delta k = m^2gd/\hbar^2k_1$.

For destructive interference we must have $l\Delta k = (2n + 1)\pi$. These last two relations, together with the original expression for k_1^2, lead to the result

$$E = \frac{m^3}{2}\left(\frac{gld}{(2n + 1)\pi\hbar}\right)^2.$$

117

We assume the rod to be long compared with all the other dimensions involved. We denote the charge per unit length by σ and the distance of the rod from the (vertical) stream of water by x (see the diagram). Then the field at B is $\sigma/2\pi\epsilon_0 r$, and the x component of this field is $E_x = \sigma x/2\pi\epsilon_0(x^2 + y^2)$. If we make the approximation that this is a *uniform* field, and ignore effects due to the y component of E, then the x component of the field inside the jet is $2E_x/(\epsilon + 1)$. (This result can be established by invoking the boundary conditions that the tangential component of E and the normal component of D must be continuous across the boundary.) The polarisation is

$$P \equiv \chi\epsilon_0 E = (\epsilon - 1)\epsilon_0 2E_x/(\epsilon + 1) \simeq 2\epsilon_0 E_x$$

since $\epsilon \gg 1$. An element of the jet of length δy will have a dipole moment $P = 2\epsilon_0 E_x A\ \delta y$, where A is the cross-sectional area at B. It will thus experience a horizontal force $P(\partial E_x/\partial x)$ and therefore an acceleration $f = P(\partial E_x/\partial x)/A\rho\,\delta y$ (ρ is the density). Upon making the various substitutions, we find

$$f = \frac{\sigma^2}{2\pi\rho\epsilon_0}\frac{x(y^2 - x^2)}{(x^2 + y^2)^3}.$$

This is clearly zero at $y = \pm x$, and has a maximum of $\sigma^2/2\pi\rho\epsilon_0 x^3$ when $y = 0$. The shape of the curve is such that it will be a fairly good approximation to set f equal to this maximum value for $x > y > -x$, and zero elsewhere. We also make the simplifying assumption that x can be regarded as a constant in the expression for f. Then if an element of the jet takes a time t_1 to fall through the range $x > y > -x$, it will acquire a horizontal velocity $\sigma^2 t_1/2\pi\rho\epsilon_0 x^3$, and, in falling for a further time t_2, will be displaced horizontally by $\sigma^2 t_1 t_2/2\pi\rho\epsilon_0 x^3 = X$, say.

From the data given, the flow velocity of the jet on emerging from the tap is $2.50\ \mathrm{m\,s^{-1}}$, giving $t_1 = 7.7 \times 10^{-3}$ s and $t_2 = 0.05$ s. We also have $\rho = 10^3\ \mathrm{kg\,m^{-3}}$, $x = 10^{-2}$ m, $X = 3 \times 10^{-2}$ m and $\epsilon_0 = 8.8 \times 10^{-12}\ \mathrm{F\,m^{-1}}$. These values give $\sigma = 2 \times 10^{-6}\ \mathrm{C\,m^{-1}}$.

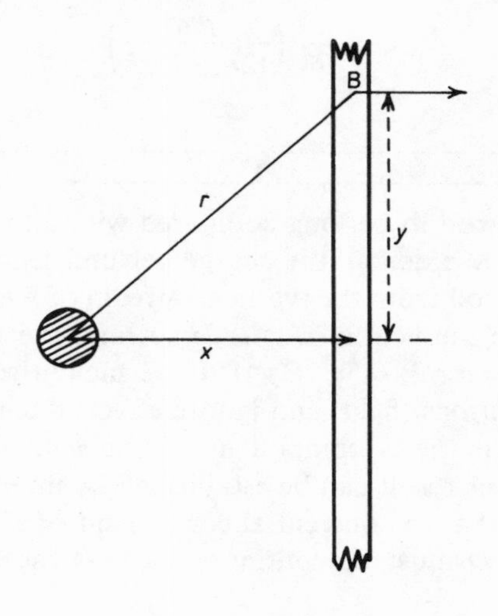

118

Of the energy $\varepsilon_1 \sigma T_1^4$ emitted per unit area by the surface at temperature T_1, a fraction ε_2 is absorbed by T_2, and thus a fraction $1 - \varepsilon_2$ is reflected. Of this, a fraction ε_1 is reabsorbed by T_1 and a fraction $1 - \varepsilon_1$ is again reflected. If we consider successive reflections in this way, and sum the resulting geometrical progression, we find that the energy emitted by T_1 and not subsequently reabsorbed (by T_1) is

$$\frac{\sigma T_1^4 \varepsilon_1 \varepsilon_2}{1 - (1 - \varepsilon_1)(1 - \varepsilon_2)}.$$

A similar expression gives the energy emitted by T_2 and not subsequently reabsorbed. The required result follows immediately.

With the intermediate screen in place (at a temperature T), the condition that, in the steady state, it loses as much energy as it gains gives $2T^4 = T_1^4 + T_2^4$. It follows at once that the net flux will be halved. The method is readily extended to show that n intermediate screens give a reduction in the heat transfer by a factor $1/(n + 1)$.

119

(i) If the real depth is s and the velocity of sound is v_s then the measured time would be $(2s/g)^{1/2} + s/v_s = t_0$, say, and the calculated depth would be $gt_0^2/2 = s_c$. Neglecting squares of small terms, we obtain $(s_c - s)/s = (2gs/v_s^2)^{1/2} = 7.2\%$.

(ii) An approximate value for the final velocity is $(2gs)^{1/2} \simeq 24\ \mathrm{m\,s^{-1}}$, showing that the Reynolds number is of the order of 10^4 for most of the time; the drag force is therefore $C_D \pi \rho a^2 v^2$, with $C_D \simeq 0.4$. The equation of motion is thus $m\ddot{s} = mg - \beta v^2$, with $\beta = C_D \pi a^2 \rho$. For the zeroth approximation, we neglect the last term, and obtain $v = gt$ if $v = 0$ at $t = 0$. For the first approximation, we insert this value in the (small) third term, to give $\ddot{s} = g - \beta g^2 t^2/m$, which at once gives $s = \frac{1}{2} gt^2 - \beta g^2 t^4/12m$ if $s = 0$ at $t = 0$. Neglect of the air drag thus introduces an error equal to the second term into the estimation of s. The fractional error is $\beta gt^2/6m \simeq \beta s/3m$. From the data given, we find that this is about 6%. Both errors make the calculated depth too large.

(iii) For the table tennis ball, the drag force is much more important relative to the weight. The velocities are smaller, but still large enough to make $R > 10^3$, i.e. C_D is still about 0.4. In this case, the terminal velocity is $(mg/\beta)^{1/2} \simeq 5.3 \text{ m s}^{-1}$. The ball will reach this speed after a few metres, so that the zeroth approximation, $s_c = \frac{1}{2}gt^2$, is a gross overestimate of the depth—by a factor of about 2.6 for $s = 30$ m. A better first approximation is to assume that it travels at its terminal velocity for the whole distance. In this approximation, the error due to the finite velocity of sound is clearly given by the ratio of the terminal velocity to the velocity of sound. The error due to the initial period of acceleration cannot be evaluated without using the exact solution of the equation of motion. This is too complicated to be evaluated in the present context. (It is, in fact, $e^{-\theta} = e^{\xi}[1 - (1 - e^{-2\xi})^{1/2}]$, where $\theta = t(\beta g/m)^{1/2}$ and $\xi = x\beta/m$ if $x = v = 0$ at $t = 0$.)

120

Since the magnetic field is everywhere perpendicular to the equatorial plane, and since the initial velocity is in that plane, the electron will always remain in that plane. Also, since the force Bev is always perpendicular to v, the speed of the electron will remain constant. If B were constant, it would thus move in a circular orbit of radius $r_0 = mv/eB$ with a periodic time $\tau_0 = 2\pi m/eB$.

However, B is not quite constant. It is smaller—and thus r will be larger—for that part of the orbit remote from the earth, and vice versa. Since we are told that $r_0 \ll R$, these variations will be small and we can approximate by assuming that the orbit consists of a succession of semicircles of radii $r_1 = mv/e(B - \delta B)$ and $r_2 = mv/e(B + \delta B)$. Thus the electron progresses a distance $2(r_1 - r_2)$ in each orbit, and this takes a time $\tau_0 = 2\pi m/eB$. Now $2(r_1 - r_2) = (4mv/e)\, \delta B/B^2$. Since the magnetic field of the earth is that of a dipole, $B \propto M/R^3$ (M is the dipole moment); hence $\delta B/B = 3\delta R/R$. It will be a fair approximation to set $\delta R = r_0/2$. Having done this we find, for the mean forward velocity, $V = 3vr_0/\pi R$. The time for a circuit of the earth is then $2\pi R^2/3vr_0 = 2\pi^2 R^2 Be/3mv^2 = \frac{2}{3}\pi^2 R^2 \omega/v^2$, as stated. A more exact treatment merely changes the value of the numerical constant.

121

For a very crude model, we suppose the water molecules to be arranged on a simple cubic lattice. Consider a cube with N atoms on each edge. Let E denote the bonding energy between nearest neighbours, and ignore all other pairs. The total energy is then

$$\tfrac{1}{2}E[Z(N-2)^3+6(Z-1)(N-2)^2+12(Z-2)(N-2)]$$

where Z is the number of nearest neighbours (= 6 in this case). The first term arises from all those molecules in the interior which have 6 neighbours. The second term arises from the molecules on the 6 faces and the third term from those on the 12 edges, which have 5 and 4 neighbours respectively. The factor of $\tfrac{1}{2}$ outside the brackets takes account of the fact that every pair will be counted twice. The expression simplifies to $3E(N^3-N^2-4)$ if $Z=6$. If a is the distance between neighbour molecules, then the volume of the cube is $V=(Na)^3$ and its surface area is $S=6(Na)^2$. The energy can thus be written

$$3E\left(\frac{V}{a^3}-\frac{S}{6a^2}-4\right).$$

The energy per unit volume, $3E/a^3$, can be associated with the latent heat H, and the energy per unit area, $E/2a^2$, with the surface energy. For the present purposes, this is equal to the surface tension T. We thus obtain $T/H=a/6$. Inserting the numerical values, we find $a=2\times10^{-10}$ m.

If we consider a sphere instead of a cube, a similar argument leads to a similar result: the energy is $\tfrac{1}{2}E(6V/a^3-S/a^2)$.

122

Since the quantity λ/r^n is to represent a force, λ must have the dimensions $ML^{n+1}T^{-2}$. We assume that η is a function of λ, m and V, but not p, as stated. Then, using the usual technique of dimensional analysis, we find that

$$\eta=\text{constant}\,(\lambda^{-2/(n-1)}m^{(n+1)/(n-1)}V^{(n+3)/(n-1)}).$$

Of the quantities on the right-hand side of this equation, only V will depend on temperature, through the equation $T\propto\tfrac{1}{2}mV^2$. Thus $\eta\propto T^{(n+3)/2(n-1)}$. Since we are told that $\eta\propto T^{0.725}$, we can solve for n and find $n=9.89\simeq10$.

123

(i) Two pulses propagate in opposite directions around the hoop. Their velocities are $c=\pm(T/\rho A)^{1/2}$, where T is the tension, ρ is the density and A is the cross-sectional area. The tension T is that needed to prevent the hoop from expanding as a result of the centrifugal force, i.e. $T = \rho A v^2$. Thus $c = \pm v$ and so one wave remains stationary in space while the other moves with a velocity $2v$ in the same direction as v.

(ii) The strain is $f = T/AE$ (E is Young's modulus). But $T = \rho A v^2$ (above), i.e. $f = \rho v^2/E$. The velocity of longitudinal waves in a material is $c = (E/\rho)^{1/2}$. Thus $f = v^2/c^2$.

(iii) No. The only forces acting on an element of length of the wire are its weight, downwards, and the reaction of the ground, upwards. The former is bigger than the latter by an amount that is just sufficient to keep the mass moving in its circular orbit. There are no longitudinal forces.

124

The temperatures and radii of the three bodies are denoted by T and R, with suffixes S, E and O to distinguish between sun, earth and sphere. If the distance from the earth to the sun is x, then $2R_S/x = 0.5° = \theta$, say. The intensity of solar radiation at the earth is

$$I = 4\pi R_S^2 \sigma T_S^4 / 4\pi x^2 = \theta^2 \sigma T_S^4/4.$$

Equating the energy received by the earth ($= \pi R_E^2 I$) with the energy radiated ($= 4\pi R_E^2 \sigma T_E^4$), we find

$$T_E^4 = \theta^2 T_S^4/16. \qquad (1)$$

With the data given, $T_E = 266$ K. If the sphere is also a black body, then, assuming that the earth subtends a solid angle of 2π at the sphere, a similar argument, taking account of radiation received from the sun and the earth, gives $T_O = 281$ K.

By adjusting the emissivity, the temperature of the sphere can be reduced, but it will still be comparable with that of the earth. Thus no adjustment of emissivities will have much effect on the interchange of energy between sphere and earth. For a high-temperature source (e.g. the sun) the energy is in the short wavelengths. A low emissivity for the sphere—and hence a high reflectivity—in this range will thus reduce its energy absorption from the sun. At long wavelengths we need a high emissivity, so that the body radiates well. Thus we can approximate to the optimum condition by neglecting the energy received from the sun altogether, and assuming $\epsilon = 1$ for both earth and sphere. This means that the energy radiated by the sphere ($=4\pi R_O^2 \sigma T_O^4$) is equal to that received from the earth ($=\pi R_O^2 \sigma T_E^4 = \pi R_O^2 \sigma \theta^2 T_S^4/16$ from (1)). Hence $T_O^4 = \theta^2 T_S^4/64$, i.e. $T_O = 190$ K, from the data given.

125

Direct substitution confirms that the expression given is a solution, and we find that

$$\exp(-\alpha\tau) = C + \alpha T. \qquad (1)$$

This cannot be solved directly for α. We therefore plot each side separately as a function of α and look for the points of intersection of the two curves.

If $C \simeq -1$, we see from the diagram that α will be very large at the point of intersection, and $\exp(-\alpha\tau) \simeq 0$. Thus, from (1), $C + \alpha T \simeq 0$, i.e. $\alpha = -C/T \simeq 1/T$. Since $N = N_0 \exp(\alpha t)$, N will increase exponentially with a time constant $T \simeq 10^{-7}$ s.

If $C \simeq +1$, the diagram indicates that it would be appropriate to approximate $\exp(-\alpha\tau)$ by $(1-\alpha\tau)$. Substituting in (1), we obtain $\alpha \simeq (1-C)/(T+\tau) \simeq (1-C)/\tau$. Thus, if $C < 1$, α is positive, and N increases exponentially with a time constant of approximately $\tau/(1-C)$, i.e. greater than 10 s. If $C > 1$, N decays towards zero, also with a time constant of approximately $\tau/(C-1)$.

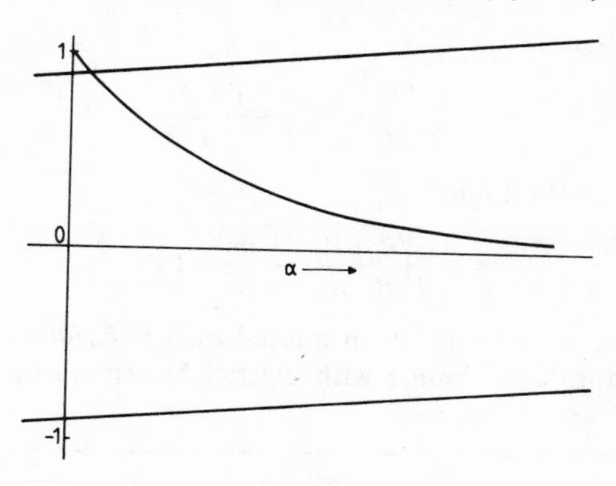

126

We assume that the field lines in the diagram are arcs of circles centred on O. Then, at a radius r, the plate separation along the field lines is $r\phi$, and the field is $V/r\phi$, where V is the potential difference between the plates. But the field at the surface of a plate is also equal to $\sigma/\epsilon\epsilon_0$, where σ is the surface charge density. Thus $\sigma = V\epsilon\epsilon_0/\phi r$. The total charge is

$$Q = \int_b^{a+b} (zV\epsilon\epsilon_0/\phi r)\, \mathrm{d}r$$

from which the stated result follows.

In fact, the field lines will be more nearly straight than was assumed above. To obtain another limiting estimate we can assume that they are quite straight—and ignore the fact that they will then not be perpendicular to the face of the conductor. A treatment along the same lines as above gives a similar result, but with ϕ replaced by $2\sin(\phi/2)$. The true value—still neglecting edge effects—will lie between the two.

127

The resistance of the wire is $R = \rho l/\pi r^2$, where ρ is the resistivity. When the longitudinal strain is equal to $\delta l/l$, the transverse strain, $\delta r/r$, will be $-\nu\, \delta l/l$. Since we have, in general,

$$\frac{\delta R}{R} = \frac{\delta\rho}{\rho} + \frac{\delta l}{l} - 2\frac{\delta r}{r}$$

it follows that

$$\frac{1}{R}\frac{\delta R}{\delta l} = \frac{1}{\rho}\frac{\delta\rho}{\delta l} + \frac{1+2\nu}{l}$$

and hence, in the limit,

$$\nu = \frac{1}{2}\left(\frac{l}{R}\frac{\partial R}{\partial l} - \frac{l}{\rho}\frac{\partial\rho}{\partial l} - 1\right).$$

This is equal to the expression quoted only if $\partial\rho/\partial l = 0$, i.e. if the resistivity does not change with elastic deformation. This is not true.

128

If a man of mass μ on earth raises his centre of mass through a height h by jumping, he has produced an energy of μgh. The energy needed for him to escape from a planet of mass m is $Gm\mu/r$, where r is the radius of the planet. The required result is obtained by equating these quantities, i.e. $gh = Gm/r$. To evaluate G, we note that for any mass x on earth, $GMx/R^2 = gx$, where M is the mass of the earth and R is its radius. This gives $h = R^2m/Mr$. If we assume that the two heavenly bodies have the same density, $m/M = (r/R)^3$. This gives $r^2 = Rh$. Setting $h \simeq 1$ m, we obtain $r \simeq 2.5$ km.

129

A circular mirror, 87 cm in diameter, would subtend the same angle at a ship 100 m away as the sun. It would thus produce an illuminated patch twice its own diameter, with the intensity rising (in a rather complicated way) from zero at the edge to a maximum in the centre equal to the incident intensity of 1 kW m^{-2}. A larger mirror would increase the size of this central patch of maximum intensity, but would not increase the flux. If reflecting the noon-day sun horizontally, it would be necessary to have an elliptical mirror with a major axis of $\sqrt{2} \times 87$ cm to allow for the obliquity. If a target patch on the ship was chosen to give normal incidence, no further obliquity factors would arise. Making an allowance for imperfect mirrors and a reflection coefficient less than unity, the incident energy flux would be somewhat reduced. N mirrors, all trained on to the same place, would increase the flux N-fold.

The temperature reached is given by balancing this input against the heat losses by radiation and conduction. It is unlikely that the timber would ignite below a 'dull red heat', about 500°C; at this temperature radiation losses would probably be the more important. The rate of radiation is then approximately 25 kW m^{-2}.

It is not possible to estimate how much should be added to this to allow for the heat losses by conduction and convection without a knowledge of the thermal constants of timber, but since wood is a rather poor conductor a factor of two would probably give a margin of safety and would also allow for the indifferent performance of the mirrors. Thus about 50 mirrors would be needed. The time taken to reach this temperature will depend on the thermal constants. For a crude estimate we can calculate the rate of heating of a surface layer 1 cm thick, assuming infinite thermal conductivity, and a density and specific heat equal, respectively, to $\frac{1}{2}$ and $\frac{1}{10}$ the corresponding values for water. The result is $d\theta/dt = P/2$, where P is the power input in kW m^{-2}. This will be an overestimate for the later stages, so the required heating time will be of the order of a minute—plenty of time for the Roman soldier to fetch a bucket of water.

130

A uniform electric field E induces a surface charge on a conducting sphere, which gives rise to a superposed (external) field equivalent to that of a dipole at the centre. The dipole moment M is $4\pi\epsilon_0 Ea^3$, where a is the radius of the sphere. Since the effective dipoles for the two spheres will have the same sign, the force between them will be repulsive. It is easily shown to be $3M^2/4\pi\epsilon_0 x^4 = 12\pi\epsilon_0 E^2 a^6/x^4$ (x is the distance between the centres). However, the induced charge on each sphere will affect the charge distribution on the other. To a second approximation, we can assume that each 'primary' dipole induces a 'secondary' dipole on the other sphere. General dimensional arguments show that the interaction between a primary and a secondary dipole is attractive and of order E^2a^9/x^7, while that between two secondary dipoles is repulsive and of order E^2a^{12}/x^{10}. Thus the total force is $12\pi\epsilon_0 E^2 a^6/x^4[1 - \alpha_1(a/x)^3 + \alpha_2(a/x)^6 + \ldots]$ where the α are numerical constants of order unity. Since a/x cannot be greater than $\frac{1}{2}$, when the spheres will be in contact, the total force will be repulsive, and not greater than the first term.

If $E = 100\ \mathrm{V\,m^{-1}}$ then, for spheres of a few centimetres radius, calculation shows that the electrostatic force will be about three orders of magnitude smaller than the gravitational force. The possibility of error due to this cause cannot, therefore, be ignored.

Magnetic fields can be treated in the same way, except that the induced dipoles will now depend on the magnetic properties of the material of the spheres.

131

We apply Boyle's Law to the air trapped in the bottle by one inversion. If a cylindrical bottle of length L contains liquid to a depth x, and the level falls to $x - \delta$ after one cycle, then we find

$$p(L - x) = [p - \rho g(x - \delta)]\,(L - x + \delta)$$

where ρ is the density of the liquid and p is the atmospheric pressure. It is convenient to express all lengths in terms of L as the unit. Then $p/\rho gL$ is large and $(\delta/L)^2$ is small. After making some judicious approximations, we obtain

$$\frac{\delta}{L}=\frac{x}{L}\left(1-\frac{x}{L}\right)\frac{\rho gL}{p}$$

i.e. small amounts emerge when the bottle is nearly full ($x \simeq L$) or nearly empty ($x \simeq 0$). The next approximation is

$$\frac{\delta}{L}=\frac{x}{L}\left(1-\frac{x}{L}\right)\frac{\rho gL}{p}\left[1-\frac{\rho gL}{p}\left(1-\frac{2x}{L}\right)\right].$$

The above argument assumes that the surface tension is large enough, and the radius of the hole small enough, for the configuration with the bottle inverted to be stable—i.e. air does not bubble upwards through the hole and permit more liquid to escape. Other effects of surface tension can be added as refinements, but the equations become very complicated. The effects will be small except when the bottle is nearly empty, since $2\gamma/r$ will be small compared with the other pressures involved.

132

The EMF induced in the ring is $E=AB\omega \sin \omega t$, where A is the area of the ring ($=\pi r^2$) and r is the radius. The energy dissipation per second is E^2/R, where R is the resistance ($=2\pi r\rho/a$; ρ is the resistivity, a is the cross-sectional area of the copper). Averaged over one cycle, the rate of energy dissipation is $A^2B^2\omega^2/2R$. This is equal to $-\mathrm{d}/\mathrm{d}t(\frac{1}{2}I\omega^2)$, where I, the moment of inertia, is equal to $\pi r^3 ad$, d being the density. These relations lead to the result

$$\frac{\mathrm{d}\omega}{\mathrm{d}t}=-\frac{A^2B^2\omega}{2IR}$$

from which we obtain

$$\omega=\omega_0 \exp(-t/\tau)$$

with

$$\tau=\frac{2IR}{A^2B^2}=\frac{4\rho d}{B^2}.$$

For the numerical values given, $\tau=1.6$ s.

133

Only those molecules which (*a*) are moving in the right direction and (*b*) have a speed greater than $2C_0$ will be present behind the piston. The fraction of all molecules which satisfy condition (*a*) is about $\omega/2\pi$, where ω is the solid angle subtended by the piston at the end of the tube. From the dimensions given, $\omega \simeq \pi \times 5^2/50^2 \simeq 1/400$. This neglects those molecules 'reflected' from the sides of the tube. If their directions are 'randomised' after reflection, the additional number will be small.

To calculate the fraction satisfying (*b*), we note that $C_0 = (\gamma p/\rho)^{1/2}$. Thus the energy of a molecule moving with speed $2C_0$ is $\frac{1}{2}m4\gamma p/\rho = 2mp\gamma V/mZ = 2\gamma k_B T = E_0$, say. ($Z$ is Avogadro's number.) The fraction with a speed greater than $2C_0$ is thus

$$\int_{E_0}^{\infty} E^{1/2} \exp(-E/k_B T)\, \mathrm{d}E \left(\int_0^{\infty} E^{1/2} \exp(-E/k_B T)\, \mathrm{d}E \right)^{-1}.$$

We can approximate to the first integral by writing it as

$$E_0^{1/2} \int_{E_0}^{\infty} \exp(-E/k_B T)\, \mathrm{d}E$$

which is equal to $k_B T E_0^{1/2} \exp(-E_0/k_B T)$. The second integral can be evaluated using the standard form

$$\int_0^{\infty} x^2 \exp(-x^2)\, \mathrm{d}x = \pi^{1/2}/4.$$

The required fraction is then $2(E_0/\pi k_B T)^{1/2} \exp(-E_0/k_B T)$ which, upon substitution of $E_0 = 2\gamma k_B T$, becomes $2(2\gamma/\pi)^{1/2} \exp(-2\gamma)$. With $\gamma \simeq 1.5$, this quantity is approximately 0.1; hence the product of the two factors (*a*) and (*b*) is 2.5×10^{-4}. The density of gas behind the piston would thus be of the order of 2.5×10^{-4} times that of a normal atmosphere.

134

The elementary kinetic theory of gases gives the thermal conductivity $\varkappa = \frac{1}{3} c\rho v \lambda$, where c is the specific heat per unit mass, ρ is the density, v is the molecular velocity and λ is the mean free path.

$c = C/M$, where C is the molar specific heat and M is the molecular weight. The simple theory gives $C = C_v$, but a better approximation is $C = C_p = \frac{5}{2}R$. Hence $c = 5R/2M = 5.19 \times 10^2\ \mathrm{J\,kg^{-1}\,K^{-1}}$.

ρ = (atomic weight)/(molar volume) = 1.78 kg m^{-3}.

v can be found from $\frac{1}{2}mv^2 = \frac{3}{2}k_BT$, i.e. $v = (3k_BT/m)^{1/2} = (3RT/M)^{1/2}$; this is equal to 4.12×10^2 m s^{-1}.

λ is given by the kinetic theory result $\pi d^2\lambda = V_m/Z$, where d is the molecular diameter, V_m is the molar volume and Z is Avogadro's number. This requires a value for d, which we find by setting the density of the solid, ρ_s, equal to $m/d^3 = M/Zd^3$. A refinement takes account of the known crystal structure of solid argon (FCC); the effect is to introduce a factor of $\sqrt{2}$ into the expression for ρ_s. We therefore obtain $d = 3.9 \times 10^{-10}$ m and hence $\lambda = 7.8 \times 10^{-8}$ m.

Combining all these results, we obtain $\varkappa = 9.9 \times 10^{-3}$ W m^{-1} s^{-1}. The accepted value is 16×10^{-3}.

135

Any correcting term must depend only on even powers of x since the situations $x = +\epsilon$ and $x = -\epsilon$ are physically identical.

The gap between the cylinders will be given by $\overline{a - b} + x \sin \theta$. If the inner cylinder is held fixed and the outer is rotated with linear speed V, the shear force on an element of area δA at position θ will be approximately $\delta A \eta V/(a - b + x \sin \theta)$. We write $\delta A = lb\ \delta\theta$, where l is the length of the cylinder. Then the total couple on the inner cylinder is

$$G = \int_0^{2\pi} \frac{b\, lb\, d\theta\, \eta V}{a - b + x \sin \theta}.$$

Since $x \ll a - b$ we can expand by the binomial theorem and then integrate. The result is

$$G = \frac{2\pi l b^2 \eta V}{a - b}\left(1 + \frac{x^2}{2(a-b)^2} + \cdots\right).$$

(This is of the right form, but it cannot be correct. The approximation that the fluid velocity varies linearly across the gap is not sufficient. It leads to different values of the couple on the inner and outer cylinders and—more importantly—it is not compatible with the equation of continuity for the flow of the fluid around the apparatus. An empirical approach is to regard the linear variation as the first term of a power series and add further terms, choosing the coefficients to satisfy the two physical conditions mentioned. The algebra becomes complicated, but the result is—for the present purposes—mainly to change the numerical coefficient in the correcting term.)

136

This is really a Hall effect, produced by the magnetic field of the current flowing in the rod itself. If we denote the current density in the rod (assumed uniform) by j then, for a rod of non-magnetic material, Ampères theorem applied to that part of the rod lying within a distance r of the axis gives, for the magnetic field B,

$$2\pi r B(r) = \mu_0 \pi r^2 j$$

i.e.

$$B(r) = \tfrac{1}{2}\mu_0 j r. \qquad (1)$$

If we denote the drift velocity of the charge carriers (electrons) by v, they will experience an inward radial force $\frac{1}{2}\mu_0 ejrv$. Charge will build up towards the axis until the resulting electrostatic field, radially outwards, just balances this force. Denoting the excess charge density per unit length within a radius r of the axis by $q(r)$, Gauss's theorem gives, for the field E,

$$2\pi r E(r) = q(r)/\epsilon_0. \qquad (2)$$

Equating the electrostatic and magnetic forces, we find

$$q(r) = \epsilon_0 \mu_0 \pi r^2 j v \qquad (3)$$

i.e. the charge density per unit length is $q(r)/\pi r^2 = \epsilon_0\mu_0 jv = \text{constant}$.

Intuitively, one might feel that the rod as a whole must remain electrically neutral and, indeed, general arguments can be adduced to show that this is so. There must, therefore, be a surface layer in which the excess charge density has the opposite sign. Continuum electromagnetic theory does not permit a calculation of the thickness of this layer, and a calculation on an atomic basis would be very difficult. However, it is clear that there is *some* tendency for charge to 'concentrate near the axis of the rod'.

If the rod carries no net charge, there will be no external electrostatic field that might be observed. If we take a model in which the central region has a uniform negative charge density (as above), while the outer region of small but macroscopic thickness has a uniform positive charge density, simple calculations show that there is also no net potential difference between the axis and the surface. However, the effect might be observed, in principle, by making a small radial hole into the centre of the rod, and inserting into it two fine wires, insulated from the rod except at their ends, which are arranged to lie at different depths. The electrons in these wires would not partake of the drift motion of those in the rod, and so would not experience the magnetic force. They would, however, detect the radial electrostatic field, so a potential difference should appear between them. To estimate its magnitude, we obtain, from (2) and (3), $E(r) = \frac{1}{2}\mu_0 rjv$. Integrating this from $r = 0$ to $r = a$ (the radius of the rod) and using the relations $I = \pi a^2 j$ for the total current and $j = nev$ (n is the conduction electron density), we find that the potential difference between centre and (almost) circumference is $V = \mu_0 I^2/4\pi^2 a^2 ne$. If we set $I = 10^3$ A and $a = 2 \times 10^{-3}$ m, then, using the values given for n and e, we obtain $V = 5 \times 10^{-7}$ V. Although small, this might be detected were it not for the large potential gradient along the rod. Under the same conditions, using a copper bar, this gradient is 0.8 V m^{-1}. This means that the two contacts required to detect 10^{-7} V would have to be placed in the same transverse plane to an accuracy of 6×10^{-4} mm in order to avoid spurious EMF. However, the situation is not quite so bad as it might appear from this, since the Hall EMF is proportional to I^2, while the ohmic drop is proportional to I. Thus if the current were reversed and mean values were taken, the problem would become less forbidding.

137

The differential equation governing heat conduction problems shows that the physical parameter involved (the thermal diffusivity) has dimensions L^2/T. Thus for two problems with geometrical similarity, and involving the same material, the time taken to reach corresponding states will be proportional to the square of the linear dimensions, i.e. to the two-thirds power of the masses involved. Thus $t = (48/2)^{2/3} \times 4$ minutes $= 33.3$ minutes.